New Synthetic Technologies in Medicinal Chemistry

RSC Drug Discovery Series

Editor-in-Chief:
Professor David Thurston, *London School of Pharmacy, UK*

Series Editors:
Dr David Fox, *Pfizer Global Research and Development, Sandwich, UK*
Professor Salvatore Guccione, *University of Catania, Italy*
Professor Ana Martinez, *Instituto de Quimica Medica-CSIC, Spain*
Dr David Rotella, *Montclair State University, USA*

Advisor to the Board:
Professor Robin Ganellin, *University College London, UK*

Titles in the Series:
   1: Metabolism, Pharmacokinetics and Toxicity of Functional Groups: Impact of Chemical Building Blocks on ADMET
   2: Emerging Drugs and Targets for Alzheimer's Disease; Volume 1: Beta-Amyloid, Tau Protein and Glucose Metabolism
   3: Emerging Drugs and Targets for Alzheimer's Disease; Volume 2: Neuronal Plasticity, Neuronal Protection and Other Miscellaneous Strategies
   4: Accounts in Drug Discovery: Case Studies in Medicinal Chemistry
   5: New Frontiers in Chemical Biology: Enabling Drug Discovery
   6: Animal Models for Neurodegenerative Disease
   7: Neurodegeneration: Metallostasis and Proteostasis
   8: G Protein-Coupled Receptors: From Structure to Function
   9: Pharmaceutical Process Development: Current Chemical and Engineering Challenges
   10: Extracellular and Intracellular Signaling
   11: New Synthetic Technologies in Medicinal Chemistry

How to obtain future titles on publication:
A standing order plan is available for this series. A standing order will bring delivery of each new volume immediately on publication.

For further information please contact:
Book Sales Department, Royal Society of Chemistry, Thomas Graham House, Science Park, Milton Road, Cambridge, CB4 0WF, UK
Telephone: +44 (0)1223 420066, Fax: +44 (0)1223 420247, Email: books@rsc.org
Visit our website at http://www.rsc.org/Shop/Books/

New Synthetic Technologies in Medicinal Chemistry

Edited by

Elizabeth Farrant
Worldwide Medicinal Chemistry, Pfizer Ltd., Sandwich, Kent, UK

RSC Publishing

RSC Drug Discovery Series No. 11

ISBN: 978-1-84973-017-4
ISSN: 2041-3203

A catalogue record for this book is available from the British Library

© Royal Society of Chemistry 2012

Published by The Royal Society of Chemistry,
Thomas Graham House, Science Park, Milton Road,
Cambridge CB4 0WF, UK

Registered Charity Number 207890

For further information see our web site at www.rsc.org

Printed in Great Britain by CPI Group (UK) Ltd, Croydon, CR0 4YY

Foreword

I think everyone recognises the pharmaceutical industry has undergone, and is still undergoing, massive changes in the way drugs are discovered, synthesised and manufactured. The medicinal chemist plays a vital role in coordinating the wide-ranging scientific disciplines and driving technological innovations in the quest for these new medicines. This enormously complex task must also be responsive to the demands of our modern society, be they for economical reasons, having enhanced safety profiles or leading to environmental issues. Similarly, time-lines and the global nature of this highly competitive business add additional burdens to the discovery process.

For these many reasons the molecular architects who design these exquisite structures and the synthesisers who transform simple building blocks to functional systems are forced to be increasingly creative and innovative by taking their craft to a higher art form. The rapid evolution and incorporation of new tools and novel technologies together with advances that arise by challenging the chemical reactivity dogmas of the past provides the engine to drive future successes.

This book refreshingly brings together diverse concepts, techniques and processes, all of which enhance our ability to assemble functional molecules and provides the reader with a modern skill set and an appreciation of the dynamic character of medicinal chemistry today. Indeed, many of the authors remove the constraints and blinkers associated with the traditional labour-intensive practices of the past and provide a glimpse of the future. The chapters reflect modern thinking in terms of automation and parallel methods of synthesis, particularly focussing on design by making what *should* be made as opposed to what *can* be made.

There is an emphasis on work-up tools using solid-supported reagents and scavengers to eliminate many of the time-consuming unit operations necessary to obtain pure materials during unoptimised synthesis sequences. These

RSC Drug Discovery Series No. 11
New Synthetic Technologies in Medicinal Chemistry
Edited by Elizabeth Farrant
© Royal Society of Chemistry 2012
Published by the Royal Society of Chemistry, www.rsc.org

concepts lead on naturally to methods of fast serial processing whereby microwave methods of heating are now commonplace in medicinal chemistry laboratories. Furthermore, opportunities arise by moving from conventional batch-mode synthesis to dynamic continuous or segmental flow-chemistry methods. This concept requires new thinking and apparatus but opens up exciting ways to conduct chemistry either in microfluidic channels or in larger systems which incorporate packed scavenger tubes to facilitate work-up using a machine assisted approach.

A further chapter focuses on high throughput reaction screening including biological methods. No longer is it acceptable to use expensive and talented operators to perform routine tasks; rather these should be relegated to more automated environments. Likewise, the use of software packages for reaction optimisation such as "design of experiment" and "principal component" analysis are now widely adopted and proving their worth in synthesis programming. The final visionary chapter on emerging technologies paints a seductive picture of the future. In particular, it features the importance of knowledge capture and its use in a closed loop, integrated and interactive fashion by bringing together wide-ranging techniques and devices.

The future is indeed a bright one and will continue to develop based upon the collective genius of its practitioners.

Steven V. Ley
Cambridge

Preface

It is fair to say that, for a synthetic chemist working in drug discovery, the last 15 years have seen sometimes uncomfortable levels of change in the tools and methods applied to the task of designing and synthesising new potential drug molecules. The experiments of the late 1990's with high throughput, almost industrialised, approaches to lead-molecule generation and testing failed to result in an associated increase of new drugs on the market. The ethos behind this movement was a response to the promise of advances in genomic technology to provide an enormous wealth of drugable targets for the industry to exploit, all needing tool molecules and lead material to start the process towards a drug. Over recent years, estimates of the number of genes that can be considered disease-modifying targets have been refined, resulting in the late Sir James Black's observation:[†]

> "The techniques have galloped ahead of the concepts. We have moved away
> from studying the complexity of the organism; from processes and organisation
> to composition."

Despite the fact that, with a few exceptions, the enormous libraries of closely related structures of the 1990's are now no longer being made, the technological ingenuity of this period has had a lasting impact on synthetic chemistry. Many of the techniques developed during this time are now being used routinely in medicinal chemistry labs the world over to increase productivity and access new chemical space; this is the true legacy of the "combichem revolution".

It is hoped that this book provides a useful background and context for scientists already engaged in drug discovery or entering this fascinating and

---

[†] *The Financial Times*, February 1st 2009, interview by Andrew Jack.

RSC Drug Discovery Series No. 11
New Synthetic Technologies in Medicinal Chemistry
Edited by Elizabeth Farrant
© Royal Society of Chemistry 2012
Published by the Royal Society of Chemistry, www.rsc.org

worthwhile profession, as well as demonstrating the undoubted benefits of the judicious use of synthetic technologies in drug discovery.

I would like to thank the chaptor authors, all of whom are experts and pioneers in these fields, for their high quality and timely contributions. In addition I acknowledge the particular contribution of Dr David Fox at Pfizer Sandwich and Gwen Jones at RSC Publishing for their "gentle" persistence in helping me get this project to completion. Special thanks also go to Rachel Osborne who was heroic in her efforts to write the chapter on microwave assisted chemistry in an incredibly short time-frame and late in the evolution of this book.

Dr Elizabeth Farrant
Director, Worldwide Medicinal Chemistry
Pfizer WRD
Sandwich, Kent, UK

Contents

RSC Drug Discovery Series No. 11
New Synthetic Technologies in Medicinal Chemistry
Edited by Elizabeth Farrant
© Royal Society of Chemistry 2012
Published by the Royal Society of Chemistry, www.rsc.org

CHAPTER 1
Introduction

ELIZABETH FARRANT

Worldwide Medicinal Chemistry, Pfizer Ltd., Ramsgate Road, Sandwich, CT13 9NJ, UK

1.1 Introduction

When I was training as a synthetic chemist just under 15 years ago, the range of technologies we were expected to become familiar with as postdoctoral chemists was very limited: we were expected to pack a perfect flash column, be adept with an inert gas/vacuum line and to be able to shim a 250 MHz NMR instrument. In some special cases we might have been required to use a HPLC. Now, a standard industrial lab is likely to be equipped with automated HPLC, flash chromatography, microwave reactors, maybe a flow reactor and parallel synthesis is expected as routine to maintain productivity. Analytically, the chemist has routine access to LC-MS, automated NMR instruments running complex experiments and open-access accurate mass determination. The synthetic chemistry laboratory has become a highly technology-enabled environment.

1.2 The Legacy of Combinatorial Chemistry

Many of the technologies now routinely used in synthesis have their roots in the combinatorial chemistry paradigm of the late 1990's. As the possibilities in drug discovery resulting from the sequencing of the human genome culminated in a rough draft announced by the Sanger Institute in 2001, the need to discover ligands for these estimate 3 000 to 10 000 potential disease genes[1] led to the implementation of bead-based combinatorial mixture libraries. Using this

RSC Drug Discovery Series No. 11
New Synthetic Technologies in Medicinal Chemistry
Edited by Elizabeth Farrant
© Royal Society of Chemistry 2012
Published by the Royal Society of Chemistry, www.rsc.org

technique, libraries of compounds of immense theoretical size could be man-ufactured but very soon it became clear that their utility was severely hampered by the deconvolution of any active products, the range of chemistry suitable for use with solid support and the close structural similarity of all the molecules generated.

The field evolved gradually into what is now practiced as high throughput medicinal chemistry, focusing on the synthesis of pure single compounds through solution-phase methods using diverse and imaginative chemistries with short cycle times from array design to biological test.

Many of the analytical and purification technologies developed during this time, including high throughput open-access LC-MS with UV and evaporative light scattering detection, mass-directed high throughput purification, auto-mated medium-pressure liquid chromatography and high throughput flow NMR are now in routine use in standard synthetic chemistry labs.

In addition, the methods developed to carry out high throughput plate-based chemistry have evolved as an approach to generating rich data sets to guide the optimisation of chemical reactions where there is an array of reactant, solvent and condition combinations. This has also been extended to applications as diverse as biotransformation screening and de-racemisation *via* chiral salt formation. The power of this approach to find optimal reaction conditions for key reactions as well as to discover and enable new synthetic transformations has only begun to be exploited.

A recent addition to the synthetic chemist's tool box has been the use of microwave energy to heat reactions. In many cases this more efficient heating method has been shown to dramatically shorten reaction times and also improve impurity profiles.

Another key innovation of the last 15 years has been the application of microfluidics, an approach that was initiated in the analytical community, to synthetic chemistry. In its true microfluidic format this technology is being explored as a methodology for combining the efficiency of combinatorial chemistry with the fast biological feedback needed to reduce the time to go from a hit molecule to a lead.[2] In addition, conducting chemistry in larger (mesofluidic) tube reactors has also grown in popularity due to the ability to improve reproducibility of heating and mixing over the standard round-bot-tomed flask. One interesting and fruitful application has been to use this approach to help control reactions using unstable intermediates and as a scale-up route for reactions that progress well due to the efficient heating observed in a microwave reactor.

All of these technology solutions have contributed to the chemist's toolbox, supplementing traditional approaches and equipment, and have revolutionised the way synthetic chemists design and carry out their syntheses. The impact they have had has not been the explosion in hits and lead molecules (and drug molecules) promised by the early vision of the combinatorial chemistry—that is still an underlying problem the industry is attempting to address on many fronts. However, it has been an enabling of the creativity of the synthetic chemist to build molecules and enter novel chemical space.

The case study of sorafenib illustrates beautifully the impact these technologies have been having in a drug discovery programme which used the true power of combinatorial chemistry to solve a problem that would have blocked progress to the discovery of an important cancer therapy.

1.3 Case Study: Sorafenib

The time scale for drug discovery programmes is frustratingly slow and attrition is high; however, the mid-2000's have seen drugs entering the market whose discovery has relied heavily on the application of these novel technologies. One example is the Bayer molecule sorafenib (Nexavar®) (Figure 1.1).

Sorafenib was the first oral multikinase inhibitor on the market and was designed to target Raf which is important in tumor signaling and vasculature. It was first approved for the treatment of advanced renal cell carcinoma in 2005. Despite extensive traditional analoguing and structure–activity relationship (SAR) generation around the 17 μM high throughput screening hit **1** (Figure 1.2), the chemists were unable to improve the IC_{50} beyond 10-fold from this hit.

A high throughput chemistry programme was initiated in parallel with the later stages of this work and among the 1000 compounds efficiently generated in this manner, chemists identified compound **4** (Figure 1.3) which had an IC_{50} of 0.54 μM. Crucially, during traditional analoguing, compounds **2** and **3** had been synthesised and proved essentially inactive. These data would normally significantly deprioritise the synthesis of **4** when made by resource intensive single compound synthesis as they indicate that compound **4**, a combination of the ringed groups, would lie outside the established SAR. In this case the chemists asserted that they would not have synthesised this compound in the normal course of the drug discovery programme.[3]

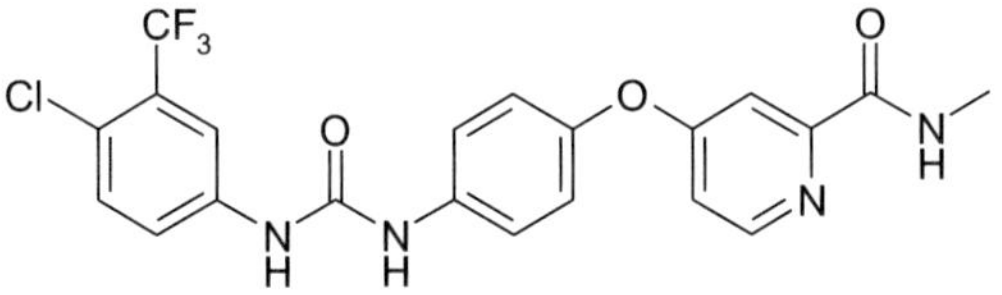

Figure 1.1 Sorafenib (Nexavar®).

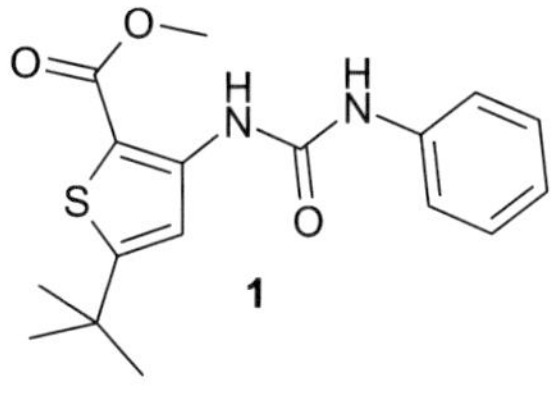

Figure 1.2 17 μM high-throughput screening hit.

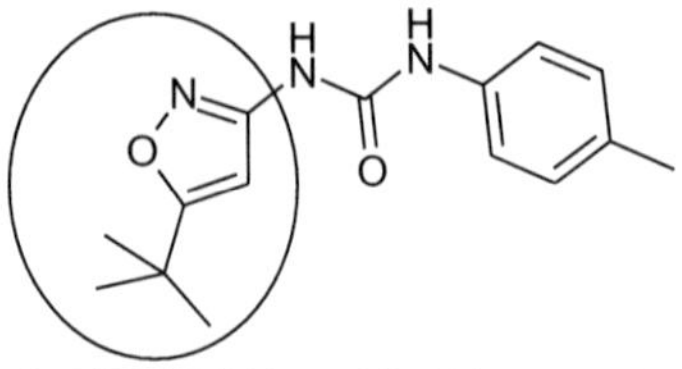

2: 34% inhibition of Raf kinase at 25uM 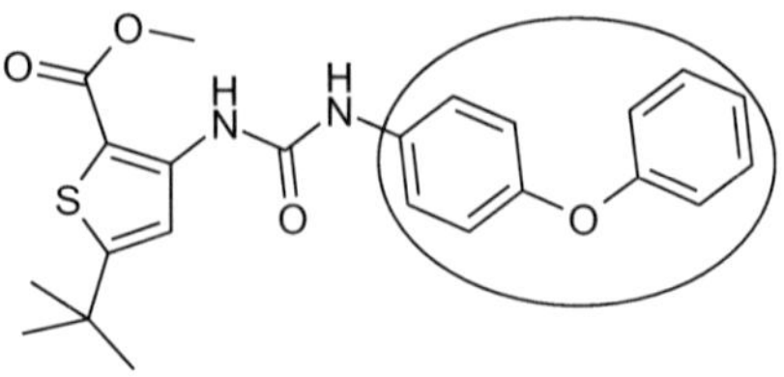   3: 34% inhibition of Raf kinase at 25uM

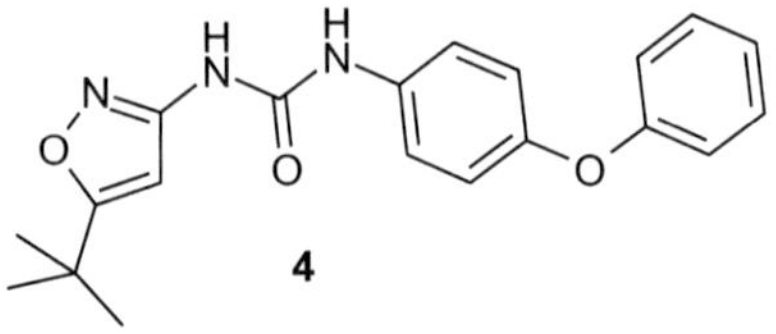

4

Figure 1.3   Key structures in the discovery of sorafenib.

Further traditional medicinal chemistry then led to the discovery of sorafenib. The researchers observed that this discovery programme shows the power of high throughput chemistry to explore efficiently the additive effects of medicinal chemistry modifications outside the normal SAR; in this case they postulate that compound **4** may adopt a binding conformation different from that of compounds **2** and **3**, explaining the divergence from the initially proposed SAR.

1.4   Conclusion

In many ways the flowering of technology development in the 1990's was largely about a wish to increase productivity in response to the increases in capacity in genomics and the promise of thousands of new drug discovery targets. In practice, however, as has often been observed, this resulted early on in an increase in the size of the drug discovery haystack rather than a rise in the number of needles found. As the following chapters of this book will demonstrate, the true result has been routine use in the synthesis lab of a range of new tools. These are used to their greatest effect when it is not merely to increase productivity by a numerical measure but to expand the access of the synthetic chemist to new chemical space which would not have been accessible by traditional approaches. The sorafenib story illustrates this in a programme that resulted in a marketed drug but the ensuing chapters will also show the many examples where wise use of technology has contributed to drug discovery, be it in target validation, medicinal chemistry design or the provision of a compound for drug discovery programmes.

References

1. J. Drews, *Nat. Biotechnol.*, 1996, **14**, 1516.
2. P. D. I. Fletcher, S. J. Haswell, E. Pombo-Villar, B. H. Warrington, P. Watts, S. Y. F. Wong and X. Zhang, *Tetrahedron*, 2002, **58**(24), 4735.
3. R. A. Smith, J. Barbarosa, C. L. Blum, M. A. Bobko, Y. V. Caringal, R. Dally, J. S. Johnson, M. E. Katz, N. Kennure, J. Kingery-Wood, W. Lee, T. B. Lowinger, J. Lyons, V. Marsh, D. H. Pogers, S. Swartz, T. Walling and H. Wild, *Bioorg. Med. Chem. Lett.*, 2001, **11**(20), 2775.

CHAPTER 2

High Throughput Chemistry in Drug Discovery

ANDY MERRITT

MRCT Centre for Therapeutics Discovery, 1–3 Burtonhole Lane, Mill Hill, London, NW7 1AD, UK

2.1 Introduction

Combinatorial chemistry (Combichem), the technique of preparing (large) numbers of compounds by common reactions in parallel using building blocks drawn from groups of molecules with a specific reactive functionality, is now an established approach with widespread application across a number of fields.[1–4] However, 20 years ago, when combinatorial chemistry was in its earliest days, pharmaceutical drug discovery research became one of the earliest adopters of the approach. Indeed, as will be discussed below, it was to do so with great commitment and expenditure. Up until the early 1990's chemistry approaches to drug discovery had remained generally consistent over a large number of years. Although there were many developments in both synthetic methodology and analytical technology that gave medicinal chemists greater tools, allowing the synthesis of more and more complex drug structures, the general concept of single compound synthesis, biological testing and subsequent design of the next target molecule was standard throughout the industry. So when chemistries[5] and technologies[6] developed in peptide chemistry came to the attention of medicinal chemists at a time when productivity and efficiency were becoming more challenged, the opportunity for a revolutionary change in drug discovery clearly presented itself. The era of 'Combichem' was about to commence.

RSC Drug Discovery Series No. 11
New Synthetic Technologies in Medicinal Chemistry
Edited by Elizabeth Farrant
© Royal Society of Chemistry 2012
Published by the Royal Society of Chemistry, www.rsc.org

The fact that this chapter is titled 'High Throughput Chemistry' rather than 'Combichem' illustrates how, in its application to drug discovery and medicinal chemistry, combinatorial chemistry has changed since those early days of the start of the 1990's, and how rocky a path that has been. Beginning with hype, when Combichem promised a solution to all lead discovery programmes through the concept of universal libraries, coupled to a predicted large scale reduction in medicinal chemistry requirements (and therefore resources), it was only after many endeavours that the realisation struck home that if you are looking for a needle, then maybe making the haystack bigger isn't always the best approach.[7] The terms combinatorial chemistry and Combichem fell from favour, associated with brute force 'industrialisation' that discarded the history and science of drug discovery with predictable lack of success (20/20 hindsight is, of course, always accurate), only to be reinvented quietly as the current practices of 'high throughput chemistry' and 'parallel lead optimisation'. Beneath the headlines, be they the positive 'great opportunities' of the 1990's or the negative 'plethora of data but where are the drugs' of the 2000's, the application of combinatorial chemistry to drug discovery has had a wide impact on medicinal chemistry research. Techniques, strategies, design and synthetic methodologies have all been developed and are in constant use in most drug discovery labs of today. The intent of this chapter is to illustrate with several recent examples how such approaches are in common use; however, to understand these it is worth first reviewing some of the key developments of combinatorial chemistry and their application to drug discovery.

It is beyond the scope of this chapter to provide anything like a comprehensive coverage of the field of combinatorial chemistry, whether its historical development or its application to current lead discovery and optimisation. There are many comprehensive reviews,[8–12] tabulated summaries,[13–25] books[26–30] and whole journals[31] which the reader may wish to consult if greater depth of understanding or historical context is desired. Instead this chapter will provide a more personal view of the key moments and developments of combinatorial chemistry in drug discovery to hopefully illustrate the many aspects of combinatorial chemistry, drawing on experience in large diversity library production and technology development,[32] technology transfer and change management in lead optimisation groups,[33] targeted small array approaches to lead discovery and lead optimisation using array technologies.[34]

The development of high throughput chemistry has also led to rapid changes in the IT infrastructure to support drug discovery, in areas as diverse as sample management,[35] process automation[36] and electronic registration;[37] however, these areas are beyond the scope of this chapter and the references supplied should be followed if more information is required.

2.2 The Potential of High Throughput Chemistry in Drug Discovery

Before considering the historical development of combinatorial chemistry and the current best practices, it is worth reviewing how drug discovery typically progresses and identify those areas where combinatorial and high throughput chemistry may have a significant impact.

As described by Hughes,[38] drug discovery can be roughly divided into 4 stages. The initial stage, that of therapeutic target definition, is a biology driven component which relies on chemical tools to help clarify mechanisms and pathways. As such, the opportunities for parallel chemistry methods to have an impact are limited, though they may be used in the identification and optimisation of tool compounds, especially if these are peptidic.[39]

Once a target has been defined then lead identification begins, often using high throughput screening approaches. The goal of this stage is to identify compounds that have significant potency against the target, such that they warrant further exploration to optimise against a wider range of drug-like parameters. In this phase the demand for large screening collections is often met in part through compound collection enhancement using parallel chemistry approaches, either driven by chemical diversity or targeted towards particular structural motifs associated with certain protein classes.[40,41]

The third stage, once a range of lead molecules has been identified, is lead optimisation, where the goal is to optimise against several parameters leading to a compound (or preferably several compounds) suitable for full pharmaceutical development. It is in this arena that the need to generate quality data on chemical series to support SARs (structure–activity relationships) can be rapidly facilitated through parallel chemistry approaches, at a scale of 10s to 100s of compounds at a time. Not only potency, but measured components for absorption, distribution, metabolism and elimination (ADME) properties, toxicity profiles, P450 enzyme profiles and selectivity profiles may also be explored using parallel approaches.[42]

The final stage of the drug discovery process, that of development into a drug, is the most costly component of the discovery process. Attrition, especially at later stages, has a significant impact on the overall costs of drug discovery and as such the quantity and quality of data generated in the earlier stages can have a significant bearing on the final quality (and therefore potential for success) of any drug through development.

2.3 The Start of Combichem in Drug Discovery

The development of miniaturised screening leading to high throughput approaches was a significant advancement of drug discovery.[43] The standardisation of the assay format into microtitre plates, initially a 96-well format, alongside the development of automated processing, radically changed the opportunity for screening to deliver new leads for drug discovery programmes. Automation of plate movement, liquid handling and plate reading processes meant that where a few 10s of compounds may have been tested in a day by manual techniques, suddenly 1000s were possible in enzyme, (membrane-bound) receptor and even whole cell assay format. Further enhanced by the miniaturisation of wells on the plates, from 96 to 384 (and subsequently 1536), high throughput screening of compound collections of 100 000s or more became clearly feasible and, when run alongside mechanism and knowledge/structure-based targeted screening approaches, provided much greater opportunities to identify novel lead series and structural classes.

As high throughput screening developed rapidly in the late 1980's and early 1990's attention was turned to the feedstock for such efforts: company compound collections. These had typically built up by a combination of 'file' compounds from previous and ongoing lead optimisation programmes and natural products, sourced either from in-house fermentation or through external acquisition of samples, be they derived from soil, microbes or plants. A compound collection of one to two hundred thousand such compounds was not atypical, but the potential for further growth through these traditional routes would always be limited. A 'traditional' medicinal chemist was likely to add no more than 40–50 compounds in any year and, perhaps even more significantly, any file collection built on past programmes would clearly only represent those chemical areas that had been of interest. Many collections were significantly populated by specific structural classes, for example β-lactams or steroids. Meanwhile natural products were often complex structures, difficult to work with in lead optimisation, and becoming harder to source with exclusivity. International treaties correctly limited the ability to source from countries without due regard to intellectual property ownership[44] and even when novel active natural products were identified, it was possible for more than one company to independently and concurrently identify the same structural series.[45,46]

So since high throughput screening presented the opportunity to screen 100 000s of compounds in a matter of days whilst collection sizes were still limited, alternative mechanisms to increase the collections were targeted. Collection sharing deals were struck between companies,[47] though in the naturally intellectual property (IP) conservative world of drug discovery these were deals tied with many restrictions. This concept was subsequently subsumed in the mergers of the 1990's[48] where the formation of combinations such as GlaxoWellcome, Smithkline Beecham, Astra Zeneca, Novartis, and Aventis, for example, provided immediate increases in corporate collection size. In addition, the acquisition of compounds from external sources was increased, from both commercial and academic sources. Commercial suppliers provided compounds that could be added to screening collections, though these were available to all companies, thus raising concern over intellectual property control, and at that time were limited to only a few suppliers of fine chemicals. Access to more varied chemistry was available though academic collaborations, and many academic groups found they could fund several aspects of their research with money from compound selling; however, a combination of structural integrity, purity, and sustainability of resupply were all potential issues for the pharmaceutical companies using this approach.

The optimum solution for companies appeared to be a combination of the above, but enhanced with an even greater component derived from a significant increase of productivity from their own chemists. Such internally derived compounds would be proprietary, exclusive and could be targeted if necessary to areas of most interest to the company concerned. Knowledge would be retained for further synthesis, follow-up and analogue work thus providing confidence downstream of any initial positive results. The rapid development of high throughput screening had demonstrated that technology and rethinking of strategies could, in combination, provide major increases in productivity, and drug companies began to consider whether this could be also true for chemistry.

Fortunately such ideas and approaches had already been developed, though not in the field of synthetic organic chemistry but rather in peptide chemistry. The technology and methodology of solid-phase chemistry had been developed by Merrifield[49] in the 1960's and subsequent automation of the approach, maximising the advantages of forcing conditions (through excess reagent) and purification (through filtering), was well developed by this time.[50] Indeed some solid-phase work with non-peptide structures had been developed by the 1970's[51] though this had not achieved widespread use in mainstream synthetic chemistry.

The ability to carry out peptide chemistry on support in parallel was demonstrated by Geysen et al.[52] with the development of polystyrene-coated pins. Using this methodology the synthesis could be carried out in spatially addressed arrays so that common steps (deprotection and activation steps for example) could be performed using bulk reagents and reaction vessels. At around the same time Furka et al.[53] were developing the approach of "split and mix" using resin beads to allow synthesis of large numbers of peptides (albeit as mixtures) in very few reactions (Figure 2.1). Houghten[54] introduced the compartmentalisation of resin

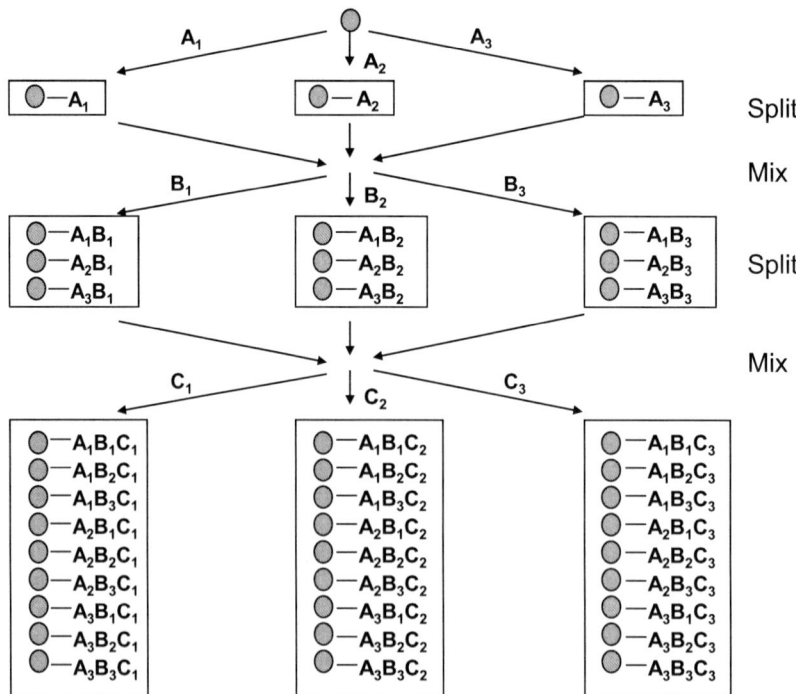

Figure 2.1   Split and mix synthesis.

beads as tea bags, thus allowing a more efficient and scaled-up handling of the process, and introducing the idea that packaged resin could then be traced through the synthetic sequence thus allowing identification of the resulting compound (or compound mixture depending on the approach adopted).

These initial developments focussed on manufacturing large numbers of small peptide fragments used, for example, to evaluate protein–protein interactions (epitope mapping)[55] or enzyme[56,57] and antibody[58] specificities. The mixtures produced using the split and mix approach needed to be deconvoluted to single active compounds, and a number of methods were developed, including iterative deconvolution[59] (fixed positions in mixtures and subsequent sub library synthesis), positional scanning[60] (replicated synthesis of same library but with a different fixed position in each mixture) and orthogonal pooling strategies[61] (replicated synthesis with orthogonal chemistries allowing different pooling strategies). The intricacies and further developments of these approaches, along with the statistical implications and subsequent results have been reviewed elsewhere.[62]

2.4 From Peptides to Small Molecules

The ability of combinatorial chemistry to make large numbers of peptides combined with various new screening approaches did not escape the attention of those involved in early hit identification programmes. Although peptides were not suitable compounds for lead identification, analysis of the drug discovery literature confirmed what many practitioners were aware of, that the large majority of drug discovery programmes involved amide bond formation or related reactions (included heterocycle formation through subsequent dehydration). As such, many of the drug discovery compounds should be accessible using similar chemistries to those of peptide synthesis.

The first 'small molecule' combinatorial library was published by Bunin and Ellman,[63] who demonstrated that a library of 40 benzodiazepines could be produced using solid-phase approaches, with three points of diversity, or variation, on the core structure (Scheme 2.1). Ellman's group expanded this work, using the pin method of Geysen to give 192 structures,[64] and further expanded this to several thousand structures in later publications.[65] De Witt described the preparation of array compounds on solid phase using the 'Diversomer' approach,[66] coupled with simple automation that was the first of many automated synthetic approaches to be introduced. That De Witt was based in industry was significant—the approach of combinatorial chemistry was clearly applicable to issues of drug discovery where obtaining data to make the next structural series decisions was the driving component of the research rather than the development of the core discipline.

Over the following few years the two main strategies of split and mix (to generate large libraries using solid-phase approaches) and parallel synthesis (focused on smaller libraries) were refined and developed. The main focus for lead discovery split and mix approaches was on a means of identifying compounds without the need for resynthesis or deconvolution stages, which

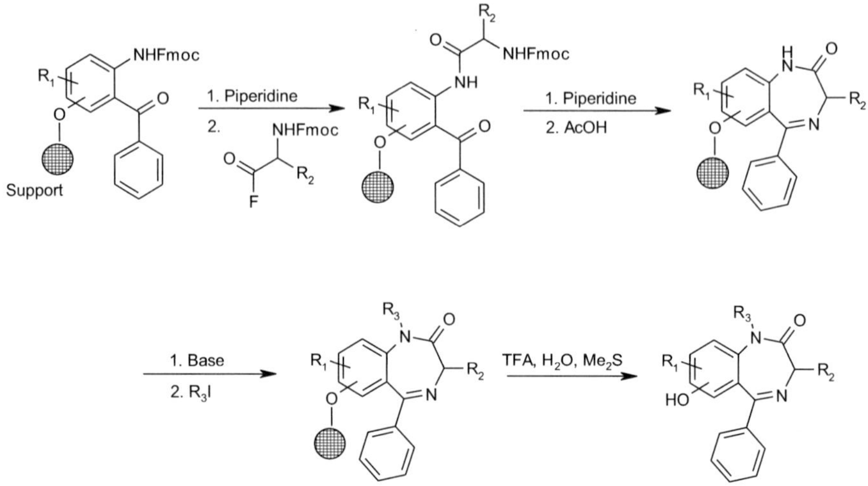

Scheme 2.1   Benzodiazepine array synthesis.

typically took too long for fast-moving lead discovery projects to allow simple mixture libraries to have an impact.[67] Tagging approaches were developed, where the solid phase was orthogonally reacted with molecules that could be 'read', typically using mass spectrometric approaches (Figure 2.2).[68] At the same time the tea bag concept of Houghten was further developed, with advancements of the container system but more importantly with the inclusion of inert radiofrequency tags.[69] These then allowed the synthetic history of any container to be either tracked or directed, thus combining the potential of split and mix with both the potential scale and single product outcome of parallel methods.

At the same time there were also rapid developments both in the range of chemistry applicable to the solid phase and in alternative approaches looking to maximise the advantages of solid-phase techniques whilst keeping those of the solution phase. The range of chemistries on the solid phase became almost as broad as traditional solution chemistry,[70–72] though in the context of this review it is worth noting (perhaps discouragingly) that even recent reviews[73] of current process chemistry activities show a similar prevalence of amide chemistry in drug programmes. Meanwhile attempts to get the solid phase 'in solution' included soluble polymers (*e.g.* polyethylene glycol monomethyl ethers,[74] non-crosslinked polystyrenes[75]) that could be precipitated for purification purposes, and the combination of fluorocarbon fluids and perfluorinated substrates[76] to allow separation from both aqueous and organic solution when required. The most applicable development to address the combination of solid- and solution-phase approaches was in supported reagents, either as scavengers to remove excess reagents or unreacted substrates[77] or as removable reagents to catalyse specific reaction steps.[78] These approaches have achieved widespread use in mainstream synthetic chemistry as

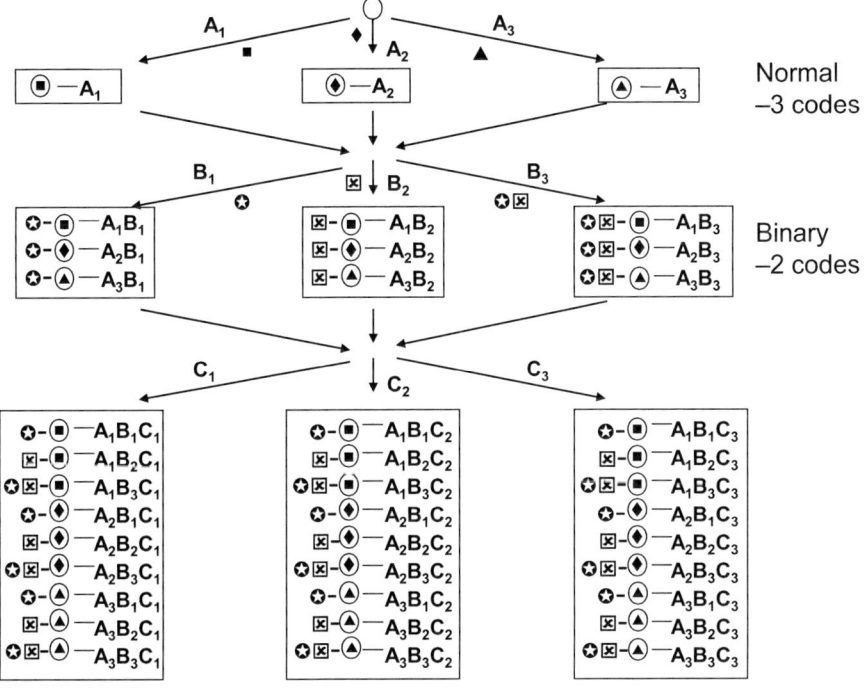

Figure 2.2 Split and mix synthesis with encoding.

well as in the combinatorial research area, and have been extensively reviewed elsewhere.[79,80]

2.5 My Library's Bigger Than Your Library: The 'Universal' Library

Before considering current best practices and the use of high throughput and parallel chemistry in drug discovery and lead optimisation, it is important to understand how the initial promise of combinatorial chemistry failed to deliver, and the subsequent backlash against large combinatorial approaches that heralded the start of the 21st century. As has been described above, high throughput screening had rapidly developed as a key component of drug discovery, to be utilised where possible alongside other lead-seeking strategies to maximise the chances of new serendipitous results. The need for 'feedstock' for the screening regime was compelling a push to maximise the scale of compound collections. New elements of diversity-driven design were exploring a whole range of new ideas on compound structures.[81–86] In this light the power of combinatorial chemistry to generate potentially millions of compounds could not be overlooked. Pharmaceutical companies rapidly followed each other in building in-house combinatorial groups, whilst external new companies were developed to focus on the technology of delivering large numbers of compounds. Many of these were subsequently acquired by pharmaceutical companies, often accompanied with the expressed intent to allow these new technology companies to continue to operate independently of the mainstream world of drug discovery.

Thus by the mid to late 1990's there were many groups using combinatorial chemistry to generate large numbers of compounds, either within pharmaceutical companies or as standalone companies operating a fee for service provision of libraries. The range of chemistry and structural motifs expanded, and groups were able to make libraries of hundred of thousands of compounds with a wide variety of structures, extremely rich in functionality.

The pinnacle of such approaches were the 'universal libraries', a concept that developed under a range of titles in many groups.[87,88] The hypothesis was a simple and powerful one. By using a set of core templates with several differentially protected functionalities and decorating these in a comprehensive combinatorial fashion with sets of compounds rich in potentially pharmacologically relevant functional groups displayed in directionally controlled manners, it should be possible to devise a single library that would cover all of 'pharmacological space' as relevant to target proteins in drug discovery. Some groups suggested this could be achieved with only a small number of core series, whilst others argued that greater central variety would be needed. However, all had one thing in common—the technology of synthesis, the concepts of spatial design of the molecules and the power of combinatorial numbers had driven the development rather than any real consideration of the nature of the resultant structures, which had to be viable structures for drug discovery optimisation programmes. Indeed, at that time the belief was expressed by some that the need for optimisation itself would be mostly eliminated—after all, from such a large and comprehensive library surely the drug itself would be present in the first screening.

2.6   From Combichem to High Throughput Chemistry: Remembering It's All About Drugs

"The pharmaceutical industry has benefited ... from rapid access to a large number of novel compounds and related biological data though combinatorial chemistry and high throughput screening. However, this plethora of data has yet to translate into clinical success."

The above extract from Oprea's review[89] of the impact of combinatorial chemistry is just one of many that could be used at this point. Clearly the generation of millions of compounds, not to mention the investment of significant resources into developing technologies, strategies and expertise had not reaped the hyped dividends so readily promised in the early days of combinatorial chemistry. So where did it go wrong?

One of the most fundamental issues was a misconception around the scale of synthetic compound numbers as they related to all of potential chemical (or biological chemistry) space. Traditional medicinal chemistry and drug discovery had been a discipline where, once biological data had pointed the direction, the next compound for test used to take a week to prepare, and a medicinal chemist was seen as prolific if they added 100 test compounds over the lifetime of a particular project. The promise of 100 000 or more compounds from a small team and a few weeks' effort was therefore clearly a step change. Multiply that by concerted planning and the promise of hits every time from a library of maybe 1–2 million compounds appeared to be a reasonable supposition. In short, the naive view was that this step up in compound productivity was bound to yield success in screening campaigns and optimisation work. However, as computational chemists had been pointing out all along, the reality of drugable chemical space was in a completely different dimension. Final numbers vary between advocates of different techniques, but certainly the number of potential compounds to fill that space can be measured in numbers vastly greater than could ever be made (indeed greater than the number of atoms in the universe).[90,91] In a conceptual world of perhaps 10^{70} potential drug molecules then 10^6 is never going to deliver every time!

Even if the design of a library meant the potential blockbuster drug compound was meant to be in the library, the possibility of it actually being present was limited by the quality of the chemistry of the early libraries and, moreover, the means of assessing whether it was in there did not exist. Although analytical (and purification) tools and capabilities have become much more powerful (*vide infra*) in the early days it was only possible to assess quality through extensive validation of the chemistry on sample sets and then build confidence by sampling a subset of final compounds, though even this step was not viable if split and mix approaches yielding mixtures of compounds were being pursued. Solid-phase methods in particular were prone to producing varied yields in parallel steps, and the final cleavage of compounds could often generate unexpected and indefinable products due to the often forcing nature of cleavage conditions.[92]

The combinatorial chemists of the 1990's set themselves up as the new force in drug discovery. Although other areas of chemistry saw and utilised the potential of combinatorial approaches[93] it was in drug discovery that the practitioners viewed their way as revolutionary, leading as it would to a complete change in approaches to lead identification. As such, those who got involved in the field were often excellent scientists who were driven by the development of technology and the strategies of maximising the value of those technologies. Attempts to spread combinatorial approaches into mainstream drug discovery were at best of limited impact.[33] The belief that they were developing a whole new, and more effective, science for drug discovery is well illustrated by the publication challenges and how they were overcome. As the early practitioners of combinatorial chemistry looked to publish work they found the mainstream journals reluctant to accept manuscripts, demanding as they did levels of quality assurance and data than were not only not being gathered but, due to the nature of the techniques of the day, were not even feasible. Rather than work within the established literature constraints to refine how combinatorial chemistry could be adapted, the result was the establishment of new journals dedicated to the science of Combichem.[31]

The separation of combinatorial technology approaches from mainstream drug discovery had a most significant impact on the design of libraries. Driven as it was by the desire to produce large numbers and to make maximum use of the associated technologies, it was almost inevitable that the libraries produced would have large, highly functionalised structures.[94] In addition the production of large numbers of compounds around similar core structures created an illusion of diversity but in reality exacerbated the issue identified so much earlier within compound collections being dominated by common core motifs.

The rehabilitation of combinatorial chemistry (as high throughput chemistry) was enabled by a number of analyses of problems identified with earlier approaches, as well as more widespread development of understanding of factors critical in limiting attrition in potential drugs across all aspects of drug discovery. Alongside a widespread realisation that high throughput approaches needed to be considered alongside other aspects of drug discovery rather than as a separate discipline, three particular aspects are worth noting briefly as they have had major impact on the design of combinatorial approaches: the physicochemical properties of drug structures and their ability to cross biological membranes; the size of lead molecules and subsequent optimisation impact; and the incorporation of experience and knowledge into targeted library approaches.

The first of these is the seminal publication of Lipinski *et al.*,[95] outlining the 'rule of 5' as a criterion to determine the likelihood that a particular compound will pass through biological membranes, and therefore have potential to act as a drug substance. Early library structures typically had a profile of properties with mean molecular weight well above the Lipinski limits of 500, and high functionality counts (especially amide bonds) that inevitably led to too high a level of both H-bond acceptors and donors.[96] Therefore screening such libraries in any lead discovery phase, or using such design templates in lead pursuit and optimisation, is fraught with developability issues and, not surprisingly, initial results from such libraries did not become successful development candidates. As all the Lipinski parameters can be calculated from compound structures it was simple to incorporate such factors into any design approach, for example using weighted penalties in a design strategy or just setting hard limits on molecular weight and other properties.

Extending the physicochemical property limitation further, Teague and colleagues from Astra Zeneca published an analysis which showed that for lead compounds these parameters needed to be even stricter,[97] as lead optimisation consistently added both molecular weight and lipophilicity to any series as it progressed towards development candidate status. On a similar note, Hann *et al.*[98] demonstrated that the success rate of lead discovery was inversely related to the complexity of the screening structures, and that for more complex designs the likelihood of finding a successful hit against a target were very low.

Finally, the application of knowledge of past success has been brought into the design of libraries, most effectively for large targeted libraries for protein family screening. One example of this is the work of Lewell, Judd and colleagues,[99] where the knowledge of known active compounds against classes of related 7-transmembrane (7-TM) structures was used to design library building block sets incorporating 'privileged' substructures. Computational algorithms looked for common feature motifs across a range of active structures, using chemically intelligent fragmentation approaches to identify real substructures that could be introduced into new designs.

2.7 Technology to Make It Happen

Alongside the development of strategies of design and selection the development of combinatorial chemistry and the subsequent movement to high throughput chemistry approaches has driven a number of technological advances. Many of these have been 'of the moment', for example a number of high level automation approaches were extremely effective in producing large numbers of compounds but now exist only in archives of scientific equipment. Others, however, have become commonplace approaches, as have many of the developments in parallel analysis and purification, initially driven by the challenge of large number synthetic approaches.

Synthetic automation is perhaps the most notable example of such short lifetime technologies. As in other sections of this review, fully comprehensive reviews of the wide range of synthetic automation equipment are available elsewhere,[100] and only illustrative examples are used here. For example, three synthetic automated technologies were in use within GlaxoWellcome in the late 1990's, all of which are now 'retired' (and indeed examples of all have been donated to the science museum in London). Initial solid-phase work was driven by 'Advanced Chemtech' ACT machines.[101] Based around liquid handling robotics, and using proprietary designed reaction blocks there were a number of designs supporting solid-phase chemistry. At the same time split and mix approaches incorporated through the acquisition of Affymax by Glaxo-Wellcome were carried out on Encoded Synthetic Library (ESL) synthesisers,[102] with automation based around the adaption of peptide synthesisers with the ability to mix and redistribute resin to reaction vessels. Finally an arm of solution-phase-based work was supported by the development of synthetic robotics on a Tecan liquid handling bed with adaption for solvent removal through gas enhanced evaporation.[103] Between these three technologies millions of compounds were synthesised during the late 1990's; however, all were to be subsequently overtaken by the development of rf encoded encapsulated resin in the IRORI system.[104] Using automated directed sorting with capacity for up to 10 000 vessels this became the workhorse of large number synthesis, but was itself superseded by IRORI development of the X-Kan,[105] with 2D bar-coding replacing the rf tag approach. In the period of only 10 years within just one company, therefore, we have seen the introduction and subsequent displacement of more than 4 separate automated synthesisers, and in reality several more systems (*e.g.* Myriad,[106] Zinnser Sophas,[107] Argonaut Trident and Quest systems[108]) were also in use during the same period, again most of which are now retired.

The type of automated synthetic equipment outlined above has typically remained the tool of the dedicated diversity chemist, with the development of expertise around synthetic automation technology, and several groups continue to develop extensions to these approaches.[109] Of much greater impact and lasting effect was the development of simpler parallel reaction equipment, many examples of which were developed in pharmaceutical laboratories and subsequently commercialised through equipment manufacturer partnerships.[103] Many examples are available and in use today, but examples include parallel tube based reaction blocks introduced by companies such as STEM,[110] allowing controlled stirring and heating of arrays of solution-based reactions at significant scale, whilst Radleys introduced equipment based on commercialising the common practices of having several reactions on a single stirrer hotplate.[111] The Carousel took advantage of the magnetic field created by a stirrer, whilst the Greenhouse allowed reactions to be carried out readily under inert conditions. For solid-phase chemistry a number of block-based clamped filter-based systems were introduced, including Bohdan Miniblocks,[112] which took advantage of a layout format identical to microtitre plates, thus facilitating subsequent transfer to assay plates.

As discussed earlier, the development of polymer-supported reagents and sequestration agents has made solution-phase approaches to parallel chemistry viable, allowing filtration and work-up approaches to be used in parallel using filtration reagent blocks. This area has recently been reviewed[113] and includes resin capture and release approaches, tagged reagents and substrates. The following examples illustrate how these approaches have been applied in library syntheses. Strohmeier and Kappe[114] used resin capture and release steps in the preparation of 1,3 thiazine libraries (Scheme 2.2). Parlow *et al.*[115] report the use of 2 different tagged reagents to support purification by removal of reagent by-products in Suzuki coupling reactions (Scheme 2.3). Wang *et al.*[116] describe the use of polymer-supported phosphines in the wide-ranging syntheses of triazolopyridines (Scheme 2.4). Perhaps the ultimate demonstration of the power and flexibility of polymer-supported reagents and reactions is in the synthetic work of the Ley group, which has produced several publications of total syntheses of natural products (Scheme 2.5)[117] as well as a number of approaches to library and array syntheses.[118]

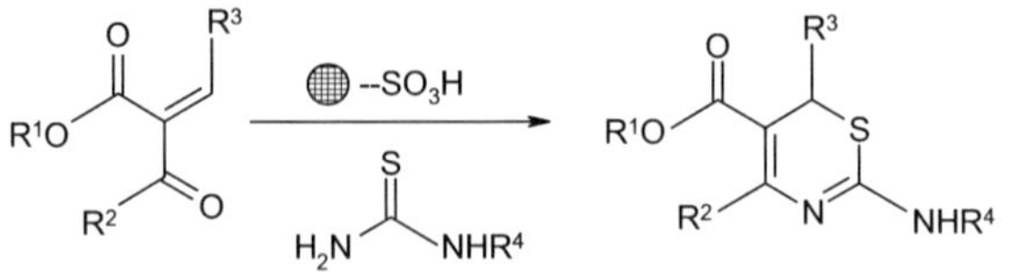

Scheme 2.2   Resin capture and release.

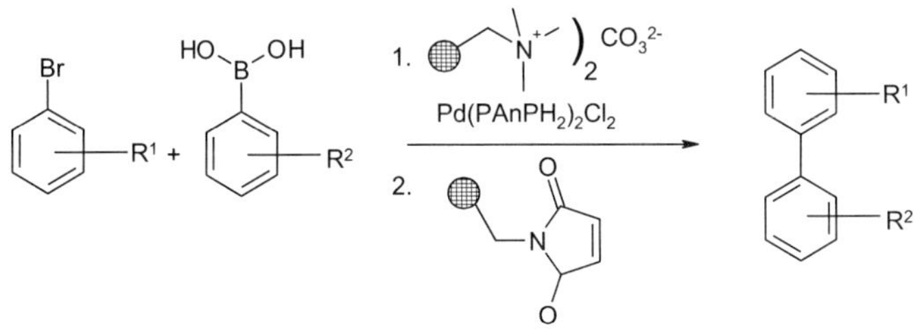

Scheme 2.3 Reagent scavenging.

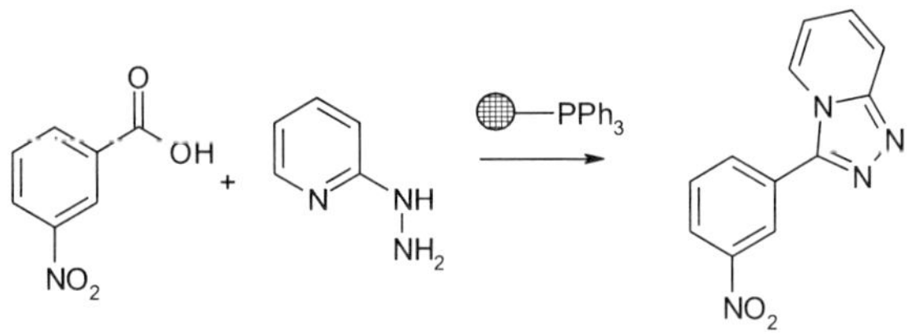

Scheme 2.4 Supported reagent.

One now commonplace technique that developed alongside the high throughput chemistry techniques has been the use of microwaves to heat and accelerate reactions.[119] Although it was initially thought was that microwaves could have a specific effect on reaction trajectories and rates, it is now generally agreed that the primary impact is the same as thermal acceleration, albeit a much faster and energy efficient one.[120] There are specific exceptions where homogeneous reactions may be affected by localised heating of solid catalysts[121] and the recent design of reaction vessels of microwave absorbing materials to maximise the effectiveness of microwave heating;[122] however, generally microwave technology has the main advantage of rapid heating, combined with being linked to automatic processing equipment which allows array chemistry to use this approach as a very specific tool for rapid compound synthesis. For example, a recent synthesis of dihydropyrimidone libraries using stepwise multi-component Biginelli chemistry and Pd/Cu mediated cross-coupling reactions, both accelerated and in high yield, illustrates some of the range and impact of microwave assisted synthesis (Scheme 2.6).[123]

Alongside parallel synthesis developments, the ability to analyse and purify a large number of compounds has also developed extensively. The use of scavenger reagents and supported sequestration approaches alongside catch and release methodologies certainly improved the purity and quality of combinatorial chemistry reactions. However, it has been the development of fast, automated LC-MS analysis systems[124] and the more recent development of fast, parallel, mass-directed preparative LC[125] that has allowed the concept of

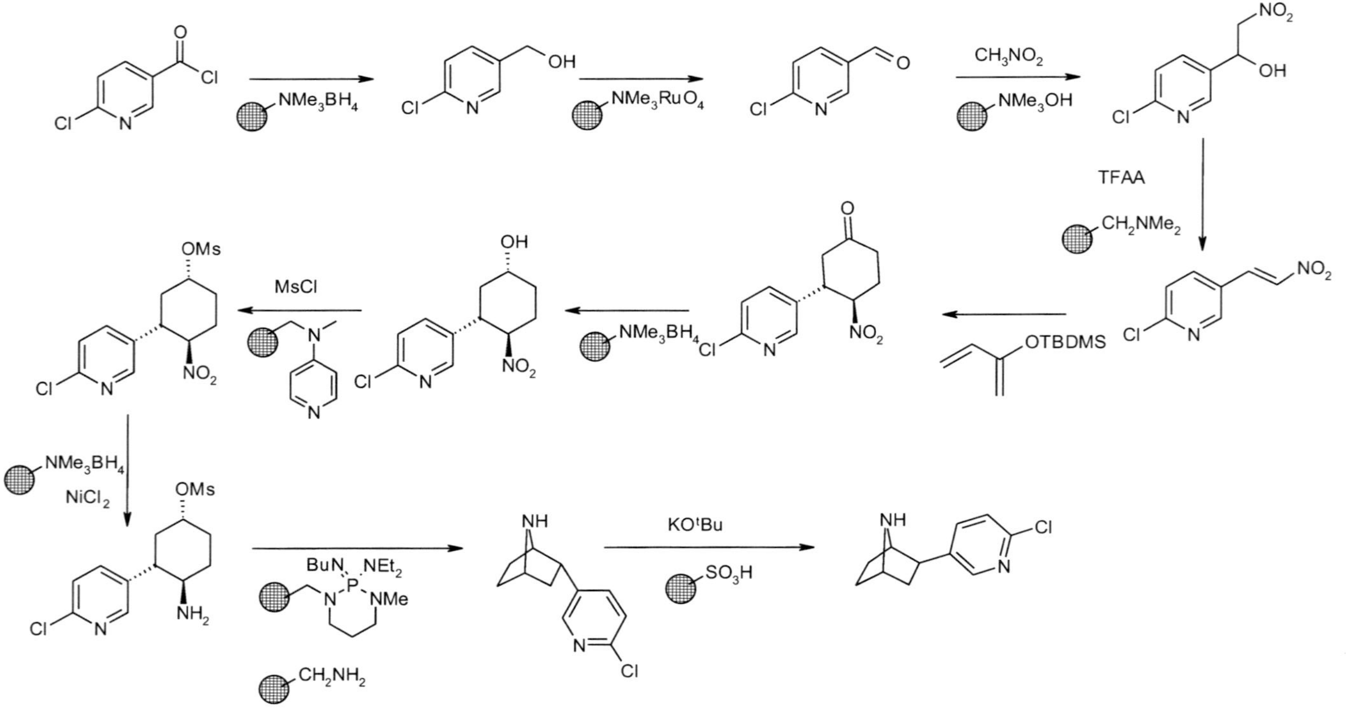

Scheme 2.5   Total synthesis of (±)epibatidine *via* polymer supported reagents.

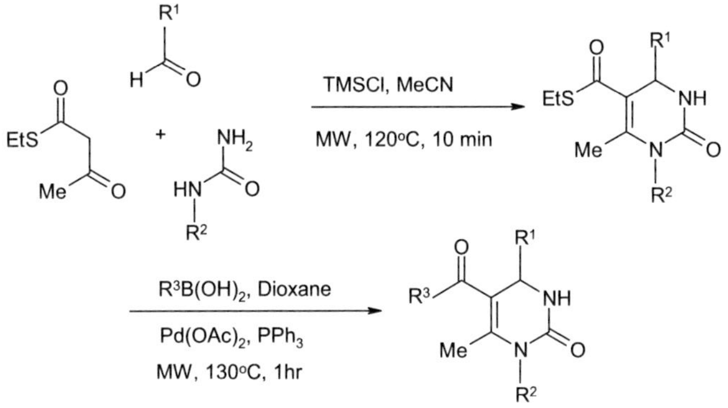

Scheme 2.6 Microwave assisted synthesis.

"purify all" to take over from previous triage processes,[126] where moderate to good purity compounds were typically processed into screening without additional purification, and only the less successful reactions were purified. The ability to estimate concentration using LC methods[127,128] has added a further level of quality into library compound in screening, as assay level concentrations can now also be determined with greater confidence rather than assuming only a single concentration across an entire collection.

2.8 Illustrative Approaches in Drug Discovery

As detailed above, high throughput chemistry has the potential to have an impact across a broad range of drug discovery stages, and is in widespread use. Any attempt to illustrate how combinatorial chemistry and parallel methods are being used must therefore be very selective, and the following examples have been drawn together to illustrate just a fraction of the possibilities. Moreover, the selection of the particular examples is not meant to identify these as standing scientifically above many other possible examples, though hopefully they illustrate well the general principles involved.

As the greatest potential impact of parallel chemistry is in the lead optimisation phase the examples start with a look at some approaches to SAR-type development of lead series. The next set looks at larger combinatorial approaches. The promise of combinatorial chemistry in driving lead discovery through high throughput screening, though somewhat tarnished by the experiences of the past few years, is still very powerful, and illustrated here with a selection of split and mix approaches. Allowing the target to define the structure of compounds in a library is the concept behind dynamic combinatorial chemistry, combining as it does the promise of fragment-based screening approaches with high throughput chemistry methods, and a number of examples are presented below. Finally combinatorial and parallel chemistry has had an impact on many other aspects of drug discovery and a few examples of such approaches are illustrated.

2.8.1 SAR Development using Parallel Chemistry

An excellent illustration of the power of parallel chemistry when used as part of a progression cascade along with high throughput screening (HTS), structural studies and structure-based design approaches is the discovery of small molecule non-chiral renin inhibitors reported by Pfizer.[129] An initial high throughput screen identified a weakly potent compound (IC_{50} 27 µM) representing a novel small molecule series. Based on this initial template 450 compounds were rapidly prepared through parallel solution-phase reductive amination chemistry, leading to compounds with single figure micromolar activity. X-ray crystallography of the protein and scaffold hopping from a previously identified series produced compounds with sub-micromolar activity, and subsequent structure-based design led to the identification of a compound with an IC_{50} of 91 nM (Scheme 2.7).

As an alternative to HTS providing the initial compounds for development by parallel chemistry, Roche described the use of virtual screening to initiate the discovery of new NPY-5 receptor antagonists, again relying on fast effective parallel chemistry to take the initial moderate lead to a highly potent series.[130] Thus initial virtual screening of the compound collection based on combining topological similarity and 3D pharmacophore approaches led to 632 compounds that were subsequently screened against the receptor, with 31 compounds having moderate to good activity, and the best of these giving an IC_{50} of 40 nM. The initial investigation of this compound was through around 100 compounds generated from a solution-phase 2-step synthesis of aminothiazoles, utilising the intermediate thiourea as a purification step, as this

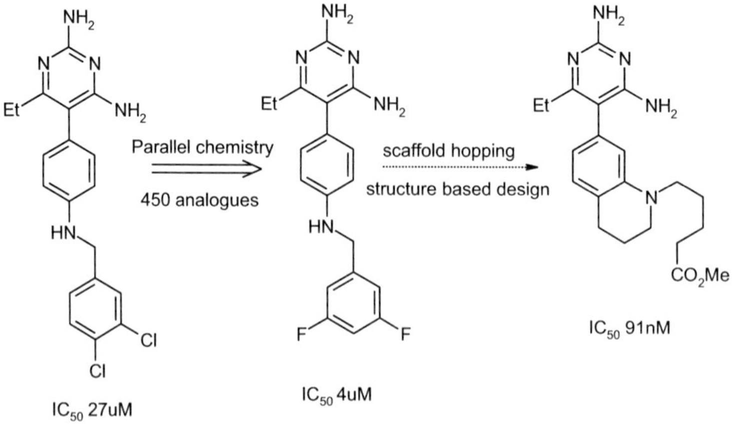

Scheme 2.7   Identification of Renin inhibitors.

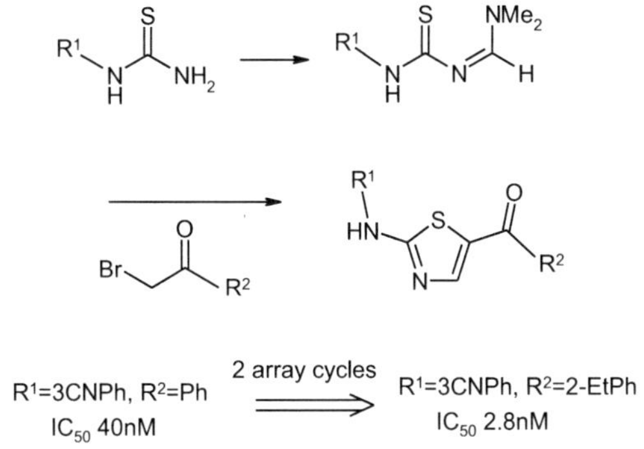

Scheme 2.8 Identification of NPY-5 receptor antagonists.

precipitated out of the synthetic reaction and was easily filtered and excess reagents washed away. Several SAR elements were identified in this initial array, and a subsequent array of 40 compounds refined the series, with the best compound having an IC_{50} of just 2.8 nM (Scheme 2.8).

One alternative to exploring lead optimisation through sequential array chemistry is to explore a molecule in parallel experiments exploring separate regions of the structure, and then combine the resulting data to lead to the optimum compound design. BMS used such an approach in the development of 2-arylbenzoxazoles as cholesterol ester transfer protein (CETP) inhibitors.[131] High throughput screening had identified a novel benzoxazole template as of interest against CETP, with an IC_{50} of 10 μM. The structure of the compound was such that exploration of either the benzoxazole ring (A in Scheme 2.9) or of the aryl ether functionality (B) was amenable to rapid parallel synthesis and two separate strategies were developed to explore these in parallel. A set of around 40 benzoxazoles was synthesised using condensation of *o*-phenolic amides intermediates whilst maintaining the *ortho*-methyl substitution of the phenolic ether in all compounds. At the same time, a series of over 50 aryl ethers was prepared using simple amide chemistry and α-keto halide displacement with phenols, in this case maintaining a dimethyl benzoxazole left hand system. Subsequent analysis of the resulting data sets, and synthesis of a small number of combined molecules yielded a compound with sub micromolar activity against CETP (Scheme 2.9).

If the previous example illustrated the use of orthogonal data generation to deliver final combined molecules, then the optimisation of S1 and S3 binding substituents in a novel series of acylguanidines targeting β-secretase activity by Wyeth illustrates the use of a true combinatorial approach.[132] Having identified the initial compound through a high throughput screen, X-ray structural studies of the compound and the target enzyme led to a design of a small combinatorial array of 156 pyrroles generated by condensation of appropriately constructed diketones with glycine. The diketones were prepared in a single step by parallel cross coupling reactions, with each half having an aryl

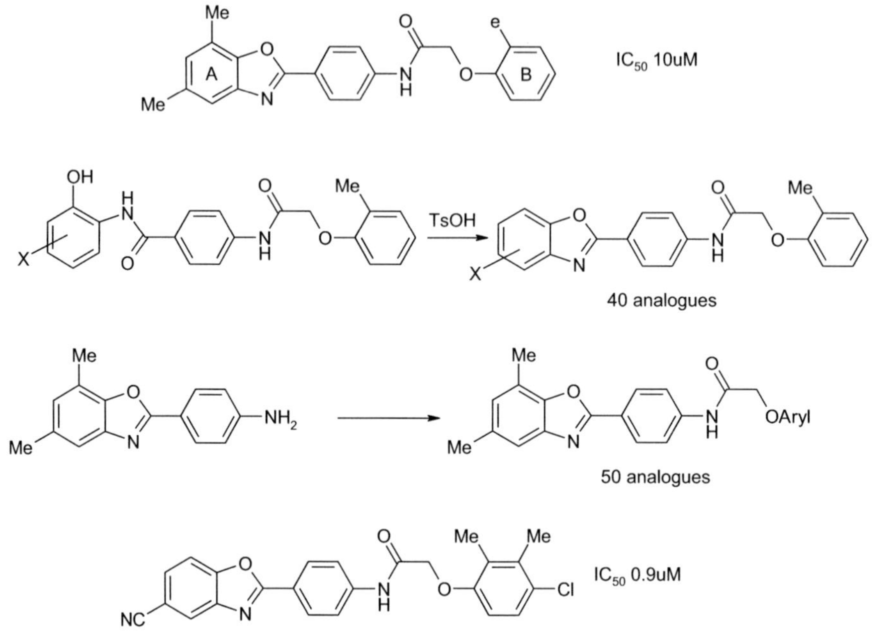

Scheme 2.9 CETP inhibitor optimisation.

substituent designed to interact either at the S1 or S3 pocket of the enzyme. 12 Ar1 and 13 Ar2 sets of compound were combined in a true combinatorial fashion to yield the final 156 compounds, with the best compound having an IC_{50} of 0.6 µM (Scheme 2.10).

Solution-phase approaches are made all the more efficient when purification approaches based on sequestration and scavenging are designed into the reaction sequence. In their optimisation of agonists for the ghrelin receptor,[133] a GlaxoSmithKline team used both polymer-supported base and solid-phase cationic extraction cartridges to maximise the purity and efficiency of a sulfonamide formation reaction. Based on an original HTS result and subsequent analogue screening, two arrays each of 24 compounds were prepared using polymer-bound *N*-methyl morpholine as the base in a sulfonamide preparation. Cationic extraction cartridges were then used to capture the required products, which all possessed a hindered piperazine moiety. Any overreacted compounds, where the hindered amine had also reacted with excess sulfonylating agents, along with excess reagents themselves were washed away before the desired pure compounds were released from the cartridge (Scheme 2.11).

An even more thorough use of scavenging approaches was described by Schering Plough,[134] where in the development of new carbamate chemistry suitable for hindered amines, two separate amine scavenging resins and a basic resin for the removal of nitrophenol were used to minimise impurities.

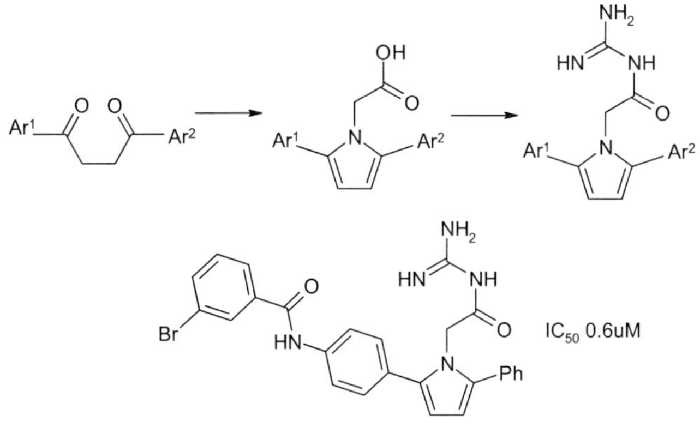

Scheme 2.10 Combinatorial optimisation of β-secretase inhibitors.

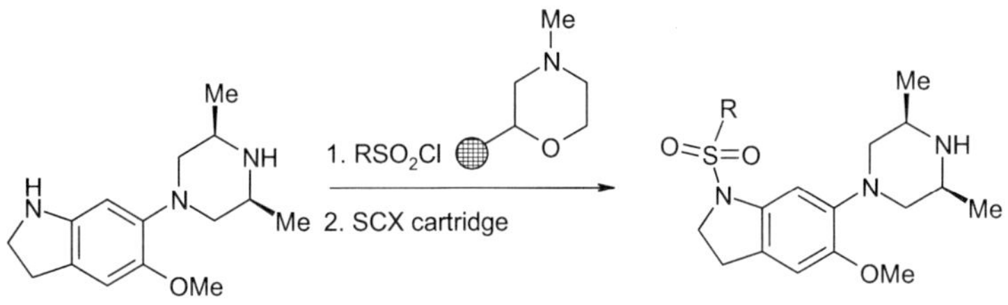

Scheme 2.11 Polymer based reagents in optimisation of ghrelin agonists.

144 compounds were prepared as potential γ-secretase inhibitors using nitro-phenylcarbonates as acylating agents for hindered amines. The removal of excess amines was facilitated using the polymer-based isocyanates or aldehydes, and the reaction biproduct nitrophenol was sequestered using basic amberlyst A26. The most potent compound thus generated had an IC$_{50}$ of 4.9 nM (Scheme 2.12).

The majority of the reported uses of parallel chemistry for SAR optimisation have been solution-phase chemistry, as it is often impractical to commit time to developing solid-phase chemistry routes. However, some solid-phase approaches have been applied to the generation of SAR. For example, Metcalf *et al.* used solid-phase syntheses and computational docking approaches to develop SAR around the hydrophobic binding pockets of the Src SH2 domain.[135] In particular, the group made use of a conserved binding motif of a primary carboxamide as a handle for solid-phase chemistry of a range of phosphory-lated aryl alkoxides (Scheme 2.13).

Click chemistry is an approach developed by Sharpless and colleagues which has been extensively used in combinatorial methodologies.[136] The application of click chemistry to capture and release approaches on solid phase have been highly effective, and in this example Prante *et al.*[137] used a REM resin[138]

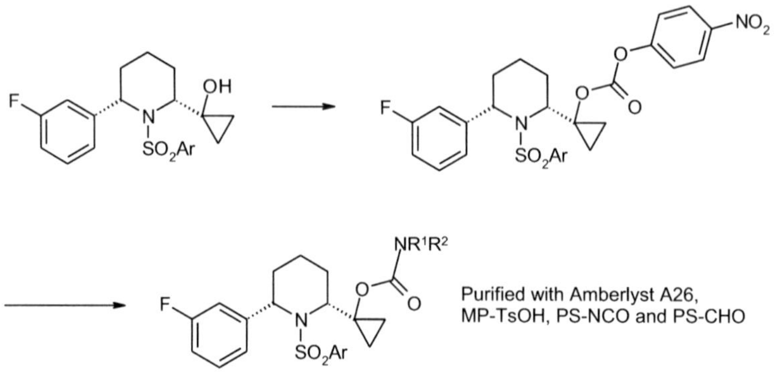

Scheme 2.12   Combined reactant and biproduct scavenging.

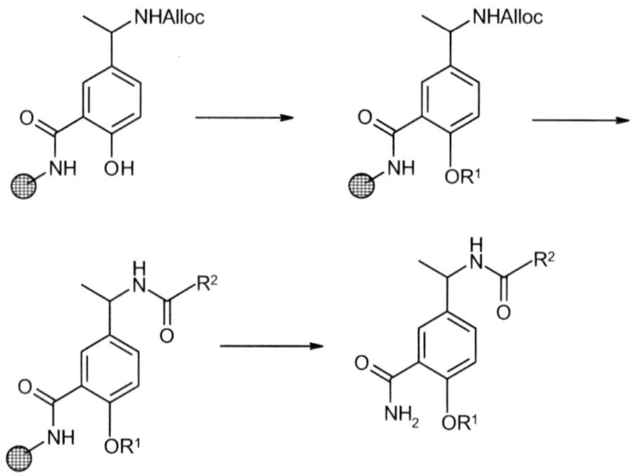

Scheme 2.13   Solid-phase synthesis in SAR development.

approach linked through click chemistry to allow the generation of 18 compounds as Dopamine D4 selective ligands for further development as PET imaging ligands (Scheme 2.14).

2.8.2   Lead Discovery: Split and Mix Examples

Although many pharmaceutical companies discarded their own approaches to large solid-phase tagged libraries through split and mix protocols, preferring either single compound synthetic protocols or rf encoding methodologies, several commercial groups have continued to use chemically encoded methods to pursue lead discovery. Pharmacopeia described a 4-component solid-phase split and mix library targeted at finding antagonists for the melanin con-centrating hormone 1 (MCH-1)[139] encoded with haloaromatic tags readable

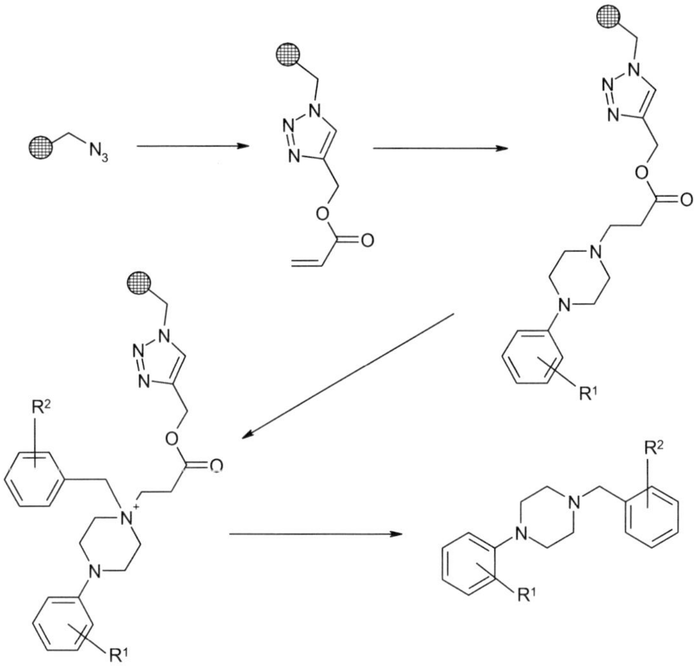

Scheme 2.14 Click chemistry applied to parallel optimisation.

through oxidative cleavage and gas chromatography, which furnished 19 470 compounds with over 95% of the library components predicted to have good adsorption properties. Several scaffolds were reductively aminated onto an aldehyde functionalised resin incorporating a photo labile linker. Subsequent amine capping reactions (amide, sulfonamide, urea) combined with differential protection of 2 amine functionalities led to the final library, which was initially screened using a small mixture (10 compound) protocol, followed by individual screening of any sub-libraries identified through the initial triage. 84 active structures were identified, the best of which had a Ki of 98 nM, which was subsequently optimised by follow up array to yield several sub-nanomolar compounds (Scheme 2.15).

Affymax have described a synthesis of a solid-phase split and mix library of over 40 000 compounds designed to identify agonists of the follicle stimulating hormone receptor (FSH receptor).[140] A thiazolidinone structure had been shown to have moderate activity from a screen against FSH, and design of the building blocks for this library incorporated a number of features from that initial hit, though many other diversity elements were also included. The library was encoded using orthogonal protection strategies and making use of quantitative encoding, whereby the encoding tags (in this case amines) varied in quantity depending on which building block was being encoded. Two active mixtures were identified on screening against FSH, which were then deconvoluted by a tiered release strategy. Thus beads from the active mixture were

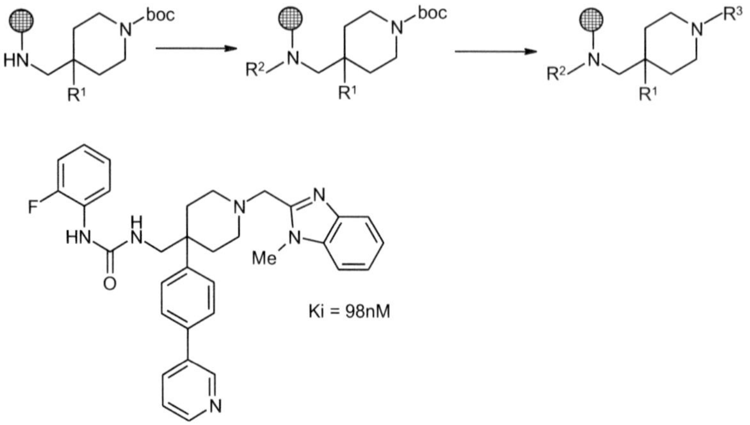

Scheme 2.15 MCH-1 antagonists *via* a tagged split and mix protocol.

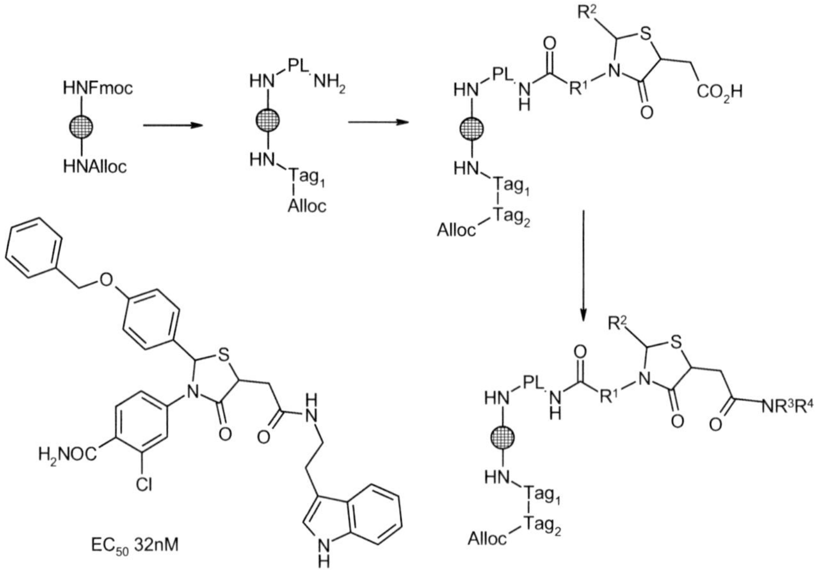

Scheme 2.16 FSH agonists *via* quantitative encoding split mix protocols.

initially distributed across plates at around 10 beads per well and photolysed for enough time to release approximately 50% of the photo-cleavable substrate. Active wells were then redistributed at 1 bead per well and the remaining material cleaved. Following this work and a subsequent round of optimisation a compound with an EC_{50} of 32 nM was identified (Scheme 2.16).

Merck have described the exploration of 2-arylindole structures as generic 7-TM targeting lead discovery libraries.[141] Fischer indole synthesis on a resin-bound aryl ketone using a safety catch linker allowed the use of amine

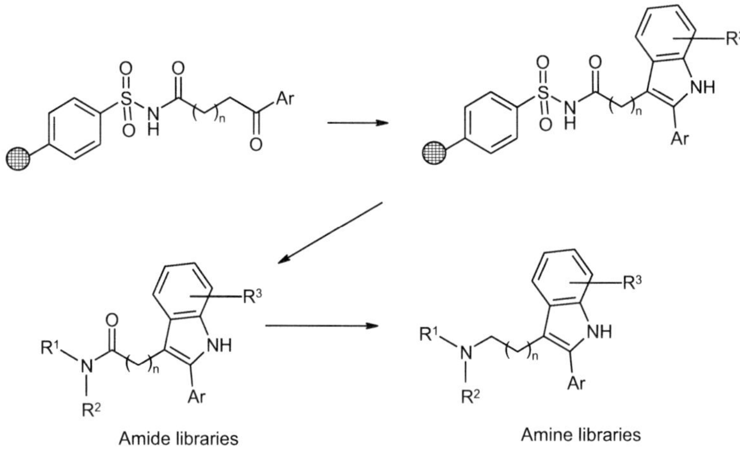

Scheme 2.17 7-TM GPCR lead discovery using safety catch methodology.

nucleophiles at the final step to release the compounds as amides (Scheme 2.17). Two 64 000 compound libraries were thus prepared, as pools each containing 400 compounds. The libraries were subsequently screening against a range of G-protein coupled receptor binding assays, and activity reported against many families, including serotonin, neurokinin and chemokine receptors. Active compounds from the pool mixtures were subsequently identified using deconvolution by synthesis of the individual members of the active pools, through a 2-stage process. Final active compounds included examples against NK1 (0.8 nM), NPY5 (0.8 nM), 5HT2a (10 nM), 5HT6 (0.7 nM) and GnRH (52 nM).

Boger *et al.* at the Scripps institute have compared the approach of positional scanning libraries with that of more traditional small mixture libraries (10 compounds per mixture, 100 mixtures) targeting paxillin/a4 inhibitors.[142] Using a simple tripeptide library, the mixture approach required one round of deconvolution to identify the best compound, with an activity of 300 nM. Positional scanning approaches with 3 sub-libraries identified the same compound without any deconvolution steps; however, as the author points out, the information-rich nature of the deconvolution approach identified substitution preferences and other subtleties that would facilitate rapid lead optimisation, data which could not be inferred from the positional scanning approach (Scheme 2.18).

2.8.3 Dynamic Combinatorial Chemistry: From Fragments to Libraries

Fragment-based screening methods have been extensively reviewed,[143,144] and their typical application to drug discovery is beyond the scope of this review. However, when fragment approaches are combined with reversible bond-forming reactions, then the resulting strategy, known as dynamic combinatorial

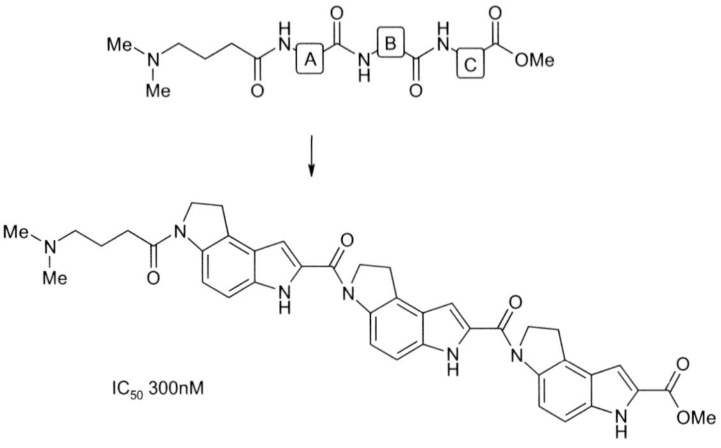

Scheme 2.18 Positional scanning approaches to paxillin/a4 inhibitors.

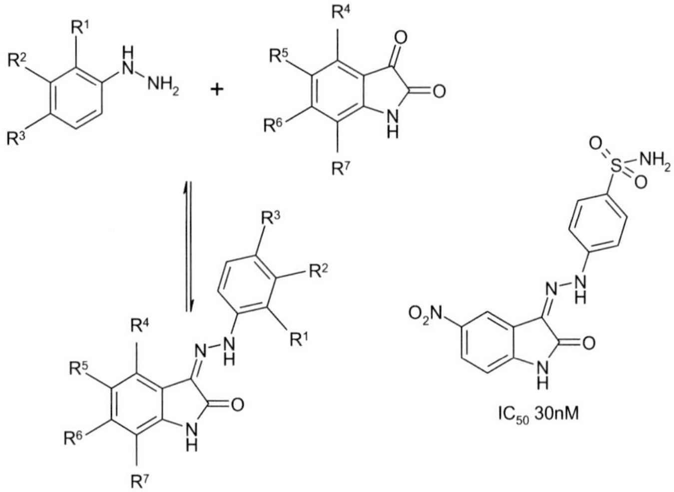

Scheme 2.19 Extended fragment screening by dynamic combinatorial chemistry.

chemistry,[145] can be a very powerful tool in lead discovery, especially when the formation of the library products can be driven in some manner by the target proteins themselves.

Astex have described the extension of fragment screening to generate larger lead-like molecules bound to CDK2.[146] Mixtures of aryl hydrazines and isatins were soaked into individual crystals of CDK2, and under equilibrating conditions reacted in a condensation reaction to form hydrazones. These were then examined *in situ* using X-ray crystallography, before the most promising compounds were resynthesised and fully profiled in typical assays. The best compound had an IC_{50} of 30 nM (Scheme 2.19).

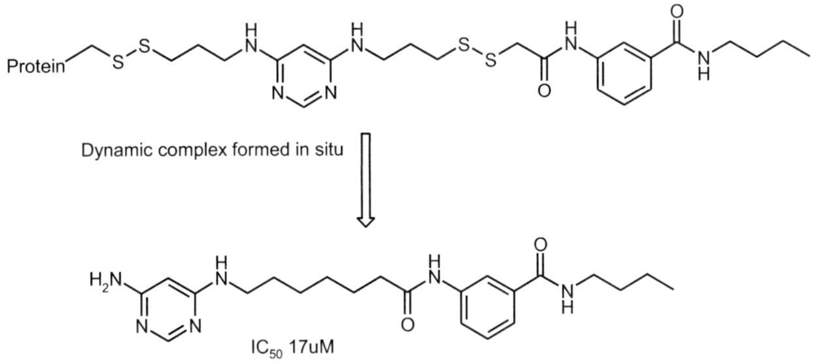

Scheme 2.20 Disulfide equilibration for dynamic combinatorial chemistry.

Another dynamic combinatorial approach using the target protein to template the chemistry was described by Sunesis pharmaceuticals.[147] In this example, the target Aurora Kinase was initially modified by site directed mutagenesis to present a cysteine SH close to the putative binding site. This handle was then exploited in a dynamic combinatorial chemistry strategy using mixtures of disulfide building blocks, which under the equilibrating conditions underwent S–S cleavage and reformation of disulfide bonds. Any building block favoured to fit in the binding site of the kinase was therefore held close to the cysteine handle in a favourable position to form a disulfide bridge. The initial series of monomers incorporated a second set of disulfide links, thus allowing a second round of equilibration with another set of monomers, and finally yielding compounds with activity in the single micromolar range (Scheme 2.20).

An alternative approach to performing reactions directly in the presence of a protein is to allow the dynamic mixture to equilibrate before introducing the target protein. Therascope have described such an approach to target novel neuraminidase inhibitors, using reductive amination chemistry on a scaffold related to known inhibitors.[148] In this example the initial imine formation was performed in the absence of the neuraminidase, and the resulting mixture reduced to yield a set of amines that could be profiled by LC-MS. The same reaction sequence was then repeated, this time with the introduction of neuraminidase during the initial imine equilibration and following the reduction step the amine profile was again analysed. A specific number of ketone examples were dramatically amplified by the addition of the enzyme, with all subsequently shown by resynthesis to be potent inhibitors of the enzyme, the most potent having a Ki of 85 nM (Scheme 2.21).

2.8.4 Other Approaches and Uses of Combinatorial Chemistry in Drug Discovery

The examples above primarily focus on the discovery and optimisation of small ('Lipinski compliant') molecules as potential oral drugs. There is, however, a

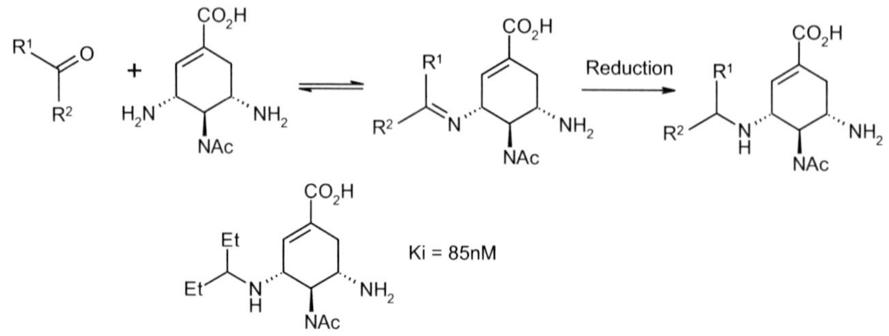

Scheme 2.21   Target protein amplification approach.

significant amount of literature describing aspects of drug discovery using parallel chemistry approaches outside of this use, and it is worth considering just a small handful of these here as illustration of the widespread potential for combinatorial approaches in drug discovery.

Antibiotic research, along with the issues of tackling multiple drug resistance strains, has become a key component of many industrial and academic research groups.[149] Although small molecule approaches are often at the forefront of such research, larger macromolecules make up a significant proportion of the current armoury of antibiotics. An approach to finding novel cyclic peptide antibiotics based on Tyrocidine A, a cyclic decapeptide, has been described,[150] using a solid-phase peptide synthesis approach and a novel efficient approach to building both the target molecules and readable tags on the bead. The 1716 analogue library was constructed on a resin-bound linker with 2 connection points, such that during synthesis of the linear peptide 90% of the material was attached *via* an allyl-protected aspartic residue whilst the other 10% was attached through an acylated residue. Treatment with triphenylphosphine and subsequent cyclisation conditions meant the 90% component cyclised and was subsequently released by a cleavage reaction, whilst the remaining 10% was retained on bead as the linear form, and could be read by amino acid sequencing techniques. Around 1% of the samples showed inhibition of growth of *Bacillus subtilis*, with the most active component having an MIC of 2 μg ml^{-1} (Scheme 2.22).

The identification of proteases as potential intervention points for disease treatment requires an understanding of the particular protease substrate.[151] Combinatorial chemistry approaches are well suited to the systematic screening of potential peptidic substrates in the search for tool compounds, which then may be used to understand the biology of the protease reactions or as screening substrates in the subsequent search for small molecule inhibitors. A recent publication describing the use of a 65 536 member octapeptide library illustrates one approach.[152] In this example the initial acylated library was exposed to the target protease whilst still resin-bound (on Tentagel resin, allowing the use of aqueous media). Visualisation of the results was through simple bead staining using amine selective reagents followed by subsequent decoding by amino acid analysis of unreacted material on the same bead. Using this method known substrate sequences for trypsin, chymotrypsin and pepsin were correctly identified (Scheme 2.23).

It is not only in the synthesis and identification of active pharmaceutical principles that combinatorial approaches can have an impact on drug discovery. For example, the use of biomolecules as therapeutics continues to offer great potential, but a major limitation remains the successful delivery of such molecules *in vivo*. RNAi therapeutics have been shown to be highly effective through the modulation of gene expression;[153] however, delivering such molecules remains challenging. Cationic lipid systems offer some potential as delivery agents, but the synthesis of these has typically been through optimised single lipid syntheses involving multiple protection deprotection steps.[154] A combinatorial library of lipid materials has now been reported, and in various screening experiments a number of the lipids were shown to enhance delivery of siRNA into HeLa cells *in vitro*, and of greater interest also demonstrated effectiveness in delivering siRNA to liver cells *in vivo* (Scheme 2.24).[155]

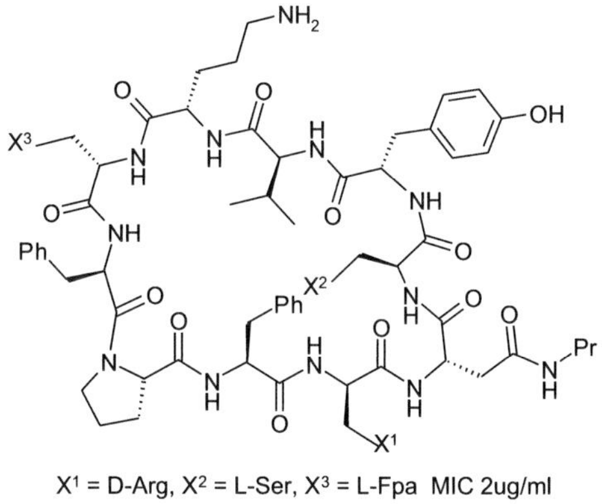

X^1 = D-Arg, X^2 = L-Ser, X^3 = L-Fpa   MIC 2ug/ml

Scheme 2.22   Cyclic peptide arrays.

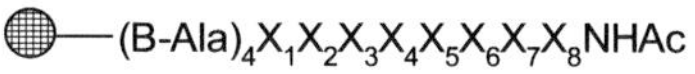

Scheme 2.23   Substrate sequence optimisation for tool compound development.

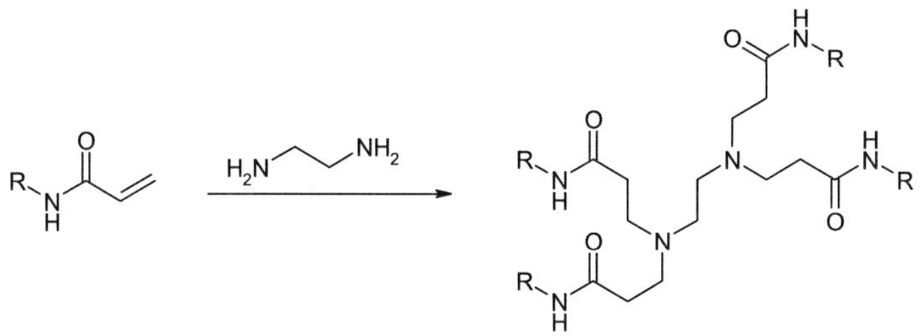

Scheme 2.24   Parallel optimisation of delivery agents.

2.9   Conclusion

Combinatorial practices, be they large library syntheses or focussed efforts of parallel chemistry around SAR generation, have become widespread throughout the drug discovery process. The initial promise of Combichem, leading as it did in the 1990's to the development of specialist teams and companies, has gone through a process of expansion, realisation, disappointment and reassessment to reach a point where it is a valuable tool, part of the overall armoury of drug discovery to be used alongside other approaches. Some aspects of the initial hype remains—there are still a number of 'combinatorial' journals for example, but whilst these focus on specialist developments and application across a wide range of fields, a casual review of any of the mainstream medicinal chemistry literature clearly identifies widespread reporting of parallel approaches as a standard approach to drug discovery and SAR development. Compound collection numbers, very much the initial driver of the combinatorial explosion, are still significant factors in defining how drug discovery can be prosecuted. However, rather than the in-house (or commissioned) combinatorial approach it is as much through purchase of compounds that these numbers are built.[156] Whereas 10 years ago purchasing compounds was very often a lottery of quality, availability and pharmaceutical relevance, it is now possible to build very large, high-quality, diverse screening sets from commercial sources. It is not that the internal industrialised approaches to screening compound synthesis epitomised by the automation facilities of GlaxoSmithKline (GSK)[157] failed in their promise, rather it is that, as in many other areas of science, the capabilities of the external world moved ahead so fast as to render the internal option much less impactful even as the true capabilities were only just being realised.

So where does that leave internal synthetic efforts? As we have seen, using parallel approaches in the evolution of SAR for lead optimisation is widespread, taking advantage of rapid turnaround of synthesis and data. Targeted screening collection development is also used, utilising privileged structures with good IP protection, and the effectiveness of this approach has been maximised by the use of outsourcing efforts to rapidly expand those collection of proprietary building blocks and scaffolds.[158] Diversity oriented synthesis, especially focused towards natural-product-like structures, is seen as a key component in increasing the drug-like potential for a compound set.[159] The use of evolutionary processes to aid cycles of design has been applied to drug discovery,[160] and if these type of approaches can be combined with application of process improvement paradigms more typical of manufacturing[161] then the efficiency and speed of lead discovery can only improve dramatically.

Combinatorial chemistry began as a tool for understanding biological processes. The application to drug discovery and the generation of small molecule drug compounds became a dream that for many developed into a nightmare of over-investment and limited return. But 20/20 hindsight is always right, and we should not be so quick to condemn the work of the earlier combinatorial pioneers. Without their pushing the boundaries of the science we would not now have an approach which, when applied correctly, can enormously shorten the discovery cycle and maximise the opportunity to optimise in parallel across a wider range of parameters than could ever have been imagined. Pick up any copy of the *Journal of Medicinal Chemistry* or *Biological and Medicinal Chemistry* and randomly open to an article—the odds are now very strong that one of the descriptors 'parallel', 'array', 'high throughput' or even 'combinatorial' will be prominent. The hype came and went but the processes embedded and stayed.

References

1. I. Takeuchi, J. Lauterbach and M. J. Fasolka, *Mater. Today*, 2005, **8**, 18.
2. A. Corma and J. M. Serra, *Catal. Today*, 2005, **107**, 3.
3. J. Scherkenbeck and S. Lindell, *Comb. Chem. High Throughput Screening*, 2005, **8**, 563.
4. D. Wong and G. Robertson, *J. Agric. Food Chem.*, 2004, **52**, 7187.
5. D. T. Elmore, *Amino Acids Pept.*, 1991, **22**, 83.
6. R. Newton and J. E. Fox, *Adv. Biotechnol. Processes*, 1988, **10**, 1.
7. R. Lahana, *Drug Discovery Today*, 1999, **4**, 447.
8. D. J. Diller, *Curr. Opin. Drug Discovery Dev.*, 2008, **11**, 346.
9. K. H. Bleicher, H.-J. Bohm, K. Muller and A. I. Alanine, *Nat. Rev. Drug Disc.*, 2003, **2**, 369.
10. G. M. Keseru and G. M. Makara, *Drug Discovery Today*, 2006, **11**, 741.
11. J. P. Kennedy, L. Williams, T. M. Bridges, R. N. Daniels, D. Weaver and C. W. Lindsley, *J. Comb. Chem.*, 2008, **10**, 345.
12. A. Lee and J. G. Breitenbucher, *Curr. Opin. Drug Discovery Dev.*, 2003, **6**, 494.
13. R. E. Dolle, B. Le Bourdonnec, A. J. Goodman, G. A. Morales, C. J. Thomas and W. Zhang, *J. Comb. Chem.*, 2008, **10**, 753.
14. R. E. Dolle, B. Le Bourdonnec, A. J. Goodman, G. A. Morales, J. M. Salvino and W. Zhang, *J. Comb. Chem.*, 2007, **9**, 855.
15. R. E. Dolle, B. Le Bourdonnec, G. A. Morales, K. J. Moriarty and J. M. Salvino, *J. Comb. Chem.*, 2006, **8**, 597.
16. R. E. Dolle, *J. Comb. Chem.*, 2005, **7**, 739.
17. R. E. Dolle, *J. Comb. Chem.*, 2004, **6**, 623.
18. R. E. Dolle, *J. Comb. Chem.*, 2003, **5**, 693.
19. R. E. Dolle, *J. Comb. Chem.*, 2002, **4**, 369.
20. R. E. Dolle, *J. Comb. Chem.*, 2001, **3**, 477.
21. R. E. Dolle, *J. Comb. Chem.*, 2000, **2**, 383.
22. R. E. Dolle and K. H. Nelson, *J. Comb. Chem.*, 1999, **1**, 235.
23. R. E. Dolle, *Mol. Diversity*, 1998, **4**, 233.
24. R. E. Dolle, *Mol. Diversity*, 1998, **3**, 199.
25. R. E. Dolle, *Mol. Diversity*, 1996, **2**, 223.
26. P. A. Bartlett and M. Entzeroth, *Exploiting Chemical Diversity for Drug Discovery*, RSC Publishing, London, 2006.

27. S. Miertus and G. Fassina, *Combinatorial Chemistry and Technology: Principles, Methods and Applications*, CRC Press LLC, Boca Raton, Florida, 2nd edn, 2005.

28. M. C. Pirrung, *Molecular Diversity and Combinatorial Chemistry: Principles and Applications., Tetrahedron Organic Chemistry Series Vol. 24*, Elsevier, Amsterdam, 2004.

29. V. Krchnak and A. Burritt, *Chemical Combinatorial Methods*, Blackwell, Oxford, 2003.

30. A. Beck-Sickinger and P. Weber, *Combinatorial Strategies in Biology and Chemistry*, Wiley and Sons, Hoboken, 2002.

31. *Journal of Combinatorial Chemistry; Combinatorial Chemistry and High Throughput Screening; Molecular Diversity.*

32. A. T. Merritt, *Comb. Chem. High Throughput Screening*, 1998, **1**, 57.

33. A. T. Merritt, *Drug Discovery Today*, 1998, **3**, 505.

34. P. W. Smith, S. L. Sollis, P. D. Howes, P. C. Cherry, I. D. Starkey, K. N. Cobley, H. Weston, J. Scicinski, A. Merritt, A. Whittington, P. Wyatt, N. Taylor, D. Green, R. Bethell, S. Madar, R. J. Fenton, P. J. Morley, T. Pateman and A. Beresford, *J. Med. Chem.*, 1998, **41**, 787.

35. J. R. Archer, *Assay Drug Dev. Technol.*, 2004, **2**, 675.

36. N. W. Hird, *Drug Discovery Today*, 1999, **4**, 265.

37. B. P. Feuston, S. J. Chakravorty, J. F. Conway, J. C. Culberson, J. Forbes, B. Kraker, P. A. Lennon, C. Lindsley, G. B. McGaughey, R. Mosley, R. P. Sheridan, M. Valenciano and S. K. Kearsley, *Curr. Top. Med. Chem.*, 2005, **5**, 773.

38. I. Hughes, in *Drug Discovery and Development, Volume 1: Drug Discovery*, ed. M. S. Chorghade, Wiley and Sons, Hoboken, 2006, p. 129.

39. S. W. Vetter and Z.-Y. Zhang, *Methods Enzymol.*, 2003, **366**, 260.

40. P. J. Eddershaw, A. P. Beresford and M. K. Bayliss, *Drug Discovery Today*, 2000, **5**, 409.

41. M. B. Brennan, *Chem. Eng. News*, 2000, **78**(23), 63.

42. D. L. Venton and C. P. Woodbury, *Chemom. Intell. Lab. Syst.*, 1999, **48**, 131.

43. G. R. Nakayama, *Curr. Opin. Drug Discovery Dev.*, 1998, **1**, 85.

44. T. D. Mays and K. D. Mazan, *J. Ethnopharmacol.*, 1996, **51**, 93.

45. J. D. Bergstrom, M. M. Kurtz, D. J. Rew, A. M. Amend, J. D. Karkas, R. G. Bostedor, V. S. Bansal, C. Dufresne, F. L. Van Middlesworth, O. D. Hensens, J. M. Liesch, D. L. Zink, K. E. Wilson, J. Onishi, J. A. Milligan, G. Bills, L. Kaplan, M. N. Omstead, R. G. Jenkins, L. Huang, M. S. Meinz, L. Quinn, R. W. Burg, Y. L. Kong, S. Mochales, M. Mojena, I. Martin, F. Pelaez, M. T. Diez and A. W. Alberts, *Proc. Natl. Acad. Sci. U. S. A.* 1993, **90**, 80.

46. A. Baxter, B. J. Fitzgerald, J. L. Hutson, A. D. McCarthy, J. M. Motteram, B. C. Ross, M. Sapra, M. A. Snowden, N. S. Watson, R. J. Williams and C. Wright, *J. Biol. Chem.*, 1992, **267**, 11705.

47. R. Thiericke, in *Modern Methods of Drug Discovery*, ed. A. Hillisch and R. Hilgenfeld, Birkhauser, Switzerland, 2003, p. 71.

48. O. Amedee-Manesmee, *Thérapie*, 1999, **54**, 419.
49. R. B. Merrifield, *J. Am. Chem. Soc.*, 1963, **85**, 2149.
50. R. B. Merrifield, *Methods Enzymol.*, 1995, **289**, 3.
51. C. C. Leznoff, *Acc. Chem. Res.*, 1978, **11**, 327.
52. H. M. Geysen, R. H. Meloen and S. J. Barteling, *Proc. Natl. Acad. Sci. U. S. A.*, 1984, **81**, 3998.
53. A. Furka, F. Sebestyen, M. Asgedom and G. Dibo, *Int. J. Pept. Protein. Res.*, 1991, **37**, 487.
54. R. A. Houghten, *Proc. Natl. Acad. Sci. U. S. A.*, 1985, **82**, 5131.
55. S. Rodda, G. Tribbick and M. Geysen in *Combinatorial peptide and non peptide libraries*, ed. G. Jung, VCH, Weinheim, 1996, p. 303.
56. P. M. St Hilaire, M. Willert, M. A. Juliano, L. Juliano and M. Meldal, *J. Comb. Chem.*, 1999, **1**, 509.
57. E. Apletalina, J. Appel, N. S. Lamango, R. A. Houghten and I. Lindberg, *J. Biol. Chem.*, 1998, **273**, 26589.
58. J. R. Appel, J. Buencamino, R. A. Houghten and C. Pinilla, *Mol. Diversity*, 1996, **2**, 29.
59. R. A. Houghten and C. T. Dooley, *Bioorg. Med. Chem. Lett.*, 1993, **3**, 405.
60. C. T. Dooley and R. A. Houghten, *Life Sci.*, 1993, **52**, 1509.
61. B. Deprez, X. Williard, L. Bourel, H. Coste, F. Hyafil and A. Tartar, *J. Am. Chem. Soc.*, 1995, **117**, 5405.
62. R. M. Kainkaryam and P. J. Woolf, *Curr. Opin. Drug Discovery Dev.*, 2009, **12**, 339.
63. B. A. Bunin and J. A. Ellman, *J. Am. Chem. Soc.*, 1992, **114**, 10997.
64. B. A. Bunin, M. J. Plunkett and J. A. Ellman, *Proc. Natl. Acad. Sci. U. S. A.*, 1994, **91**, 4708.
65. C. G. Boojamra, K. M. Burow, L. A. Thompson and J. A. Ellman, *J. Org. Chem.*, 1997, **62**, 1240.
66. S. H. De Witt, J. S. Kiely, C. J. Stankovic, M. C. Schroeder, D. M. R. Cody and M. R. Pavia, *Proc. Natl. Acad. Sci. U. S. A.*, 1993, **90**, 6909.
67. C. Barnes and S. Balasubramanian, *Curr. Opin. Chem. Biol.*, 2000, **4**, 346.
68. M. H. J. Ohlmeyer, R. N. Swanson, L. W. Dillard, J. C. Reader, G. Asouline, R. Kobayashi, M. Wigler and W. C. Still, *Proc. Natl. Acad. Sci. U. S. A.*, 1993, **90**, 10922.
69. K. C. Nicolaou, X.-Y. Xiao, Z. Parandoosh, A. Senyei and M. P. Nova, *Angew. Chem., Int. Ed. Engl.*, 1995, **34**, 2289.
70. P. H. H. Hermkens, H. C. J. Ottenheijm and D. Rees, *Tetrahedron*, 1996, **52**, 4527.
71. P. H. H. Hermkens, H. C. J. Ottenheijm and D. C. Rees, *Tetrahedron*, 1997, **53**, 5643.
72. S. Booth, P. H. H. Hermkens, H. C. J. Ottenheijm and D. C. Rees, *Tetrahedron*, 1998, **54**, 15385.
73. J. S. Carey, D. Laffan, C. Thomson and M. T. Williams, *Org. Biomol. Chem.*, 2006, **4**, 2337.

74. H. Han, M. M. Wolfe, S. Brenner and K. D. Janda, *Proc. Natl. Acad. Sci. U. S. A.*, 1995, **92**, 6419.
75. S. Chen and K. D. Janda, *J. Am. Chem. Soc.*, 1997, **119**, 8724.
76. A. Studer, S. Hadida, R. Ferritto, S.-Y. Kim, P. Jeger, P. Wipf and D. P. Curran, *Science*, 1997, **275**, 823.
77. S. W. Kaldor, M. G. Siegel, J. E. Fritz, B. A. Dressman and P. J. Hahn, *Tetrahedron Lett.*, 1996, **37**, 7193.
78. S. V. Ley, O. Schucht, A. W. Thomas and P. J. Murray, *J. Chem. Soc., Perkin Trans. 1*, 1999, 1251.
79. A. Solinas and M. Taddei, *Synthesis*, 2007, 2409.
80. S. V. Ley, I. R. Baxendale and R. M. Myers, in *Comprehensive Medicinal Chemistry 2*, ed. J. B. Taylor and D. J. Triggle, Elsevier, Amsterdam, 2006, **vol. 3**, p. 791.
81. E. J. Martin, J. M. Blaney, M. A. Siani, D. C. Spellmeyer, A. K. Wong and W. H. Moos, *J. Med. Chem.*, 1995, **38**, 1431.
82. J. H. Van Drie and M. S. Lajiness, *Drug Discovery Today*, 1998, **3**, 274.
83. J. S. Mason and M. A. Hermsmeier, *Curr. Opin. Chem. Biol.*, 1999, **3**, 342.
84. D. H. Drewry and S. S. Young, *Chemom. Intell. Lab. Syst.*, 1999, **48**, 1.
85. D. K. Agrafiotis, J. C. Myslik and F. R. Salemme, *Mol. Diversity*, 1999, **4**, 1.
86. A. R. Leach and M. M. Hann, *Drug Discovery Today*, 2000, **5**, 326.
87. M. R. Pavia, S. P. Hollinshead, H. V. Meyers and S. P. Hall, *Chimia*, 1997, **51**, 826.
88. M. J. Sofia, R. Hunter, T. Y. Chan, A. Vaughan, R. Dulina, H. Wang and D. Gange, *J. Org. Chem.*, 1998, **63**, 2802.
89. T. I. Oprea, *Curr. Opin. Chem. Biol.*, 2002, **6**, 384.
90. A. W. Czarnik, *Chemtracts: Org. Chem.*, 1995, **8**, 13.
91. Y. C. Martin, *Perspect. Drug Discovery Des.*, 1997, **7**, 159.
92. F. Z. Dorwald, *Organic Synthesis on Solid Phase*, Wiley and Sons, Hoboken, 2002.
93. A. Merritt, in *Rodd's Chemistry of Carbon Compounds, Topical Volumes, Asymmetric Catalysis*, ed. M Sainsbury, Elsevier, Amsterdam, 2nd edn, 2001, **vol. 5**, p. 259.
94. J. Alper, *Science*, 1994, **264**, 1399.
95. C. A. Lipinski, F. Lombardo, B. W. Dominy and P. J. Feeney, *Adv. Drug Delivery Rev.*, 1997, **23**, 3.
96. R. A. Fecik, K. E. Frank, E. J. Gentry, S. R. Menon, L. A. Mitscher and H. Telikepalli, *Med. Res. Rev.*, 1998, **18**, 149.
97. S. J. Teague, A. M. Davis, P. D. Leeson and T. Oprea, *Angew. Chem., Int. Ed.*, 1999, **38**, 3743.
98. M. M. Hann, A. R. Leach and G. Harper, *J. Chem. Inf. Comput. Sci.*, 2001, **41**, 856.
99. X. Q. Lewell, D. B. Judd, S. P. Watson and M. M. Hann, *J. Chem. Inf. Comput. Sci.*, 1998, **38**, 511.
100. http://www.combinatorial.com/.
101. http://www.peptide.com/.

102. B. Evans, A. Pipe, L. Clark and M. Banks, *Bioorg. Med. Chem. Lett.*, 2001, **11**, 1297.

103. N. Bailey, A. W. J. Cooper, M. J. Deal, A. W. Dean, A. L. Gore, M. C. Hawes, D. B. Judd, A. T. Merritt, R. Storer, S. Travers and S. P. Watson, *Chimia*, 1997, **51**, 832.

104. X.-Y. Xiao, R. Li, H. Zhuang, B. Ewing, K. Karunaratne, J. Lillig, R. Brown and K. C. Nicolaou, *Biotechnol. Bioeng.*, 2000, **71**, 44.

105. http://www.nexusbio.com/.

106. N. Hird and B. MacLachlan, in *Laboratory Automation in the Chemical Industries*, ed. D. G. Cork and T. Sugawara, Marcel Dekker, New York, 2002, p. 1.

107. http://www.zinsser-analytic.com/.

108. http://www.combichemlab.com/website/files/Combichem/Workstations/argonaut.htm.

109. N. Kuroda, N. Hird and D. G. Cork, *J. Comb. Chem.*, 2006, **8**, 505.

110. http://www.electrothermal.com.

111. http://www.radleys.com/.

112. http://uk.mt.com/gb/en/home/products/L1_AutochemProducts/L3_Post-Synthesis-Work-up.html.

113. J. J. Parlow, *Curr. Opin. Drug Discovery Dev.*, 2005, **8**, 757.

114. G. A. Strohmeier and C. O. Kappe, *Angew. Chem., Int. Ed.*, 2004, **43**, 621.

115. P. Lan, D. Berta, J. A. Porco, M. S. South and J. J. Parlow, *J. Org. Chem.*, 2003, **68**, 9678.

116. Y. Wang, K. Sarris, D. R. Sauer and S. W. Djuric, *Tetrahedron Lett.*, 2007, **48**, 2237.

117. J. Habermann, S. V. Ley and J. S. Scott, *J. Chem. Soc., Perkin Trans. 1*, 1999, 1253.

118. S. V. Ley, M. Ladlow and E. Vickerstaffe, in *Exploiting Chemical Diversity for Drug Discovery*, ed. P. A. Bartlett and M. Entzeroth, RSC Publishing, London, 2006, p. 3.

119. C. O. Kappe and D. Dallinger, *Nat. Rev. Drug Discovery*, 2006, **5**, 51.

120. A. De La Hoz, A. Diaz-Ortiz and A. Moreno, *Chem. Soc. Rev.*, 2005, **34**, 164.

121. B. Desai and C. O. Kappe, *Top. Curr. Chem.*, 2004, **242**, 177.

122. J. M. Kremsner and C. O. Kappe, *J. Org. Chem.*, 2006, **71**, 4651.

123. L. Pisani, H. Prokopcova, J. M. Kremsner and C. O. Kappe, *J. Comb. Chem.*, 2007, **9**, 415.

124. J. N. Kyranos, H. Lee, W. K. Goetzinger and L. Y. T. Li, *J. Comb. Chem.*, 2004, **6**, 796.

125. W. Leister, K. Strauss, D. Wisnoski, Z. Zhao and C. Lindsley, *J. Comb. Chem.*, 2003, **5**, 322.

126. S. J. Lane, D. S. Eggleston, K. A. Brinded, J. C. Hollerton, N. L. Taylor and S. A. Readshaw, *Drug Discovery Today*, 2006, **11**, 267.

127. A. W. Squibb, M. R. Taylor, B. L. Parnas, G. Williams, R. Girdler, P. Waghorn, A. G. Wright and F. S. Pullen, *J. Chromatogr., A*, 2008, **1189**, 101.

128. S. Lane, B. Boughtflower, I. Mutton, C. Paterson, D. Farrant, N. Taylor, Z. Blaxill, C. Carmody and P. Borman, *LC-GC Europe*, 2006, **19**, 161.

129. D. D. Holsworth, M. Jalaie, T. Belliotti, C. Cai, W. Collard, S. Ferreira, N. A. Powell, M. Stier, E. Zhang, P. McConnell, I. Mochalkin, M. J. Ryan, J. Bryant, T. Li, A. Kasani, R. Subedi, S. N. Maiti and J. J. Edmunds, *Bioorg. Med. Chem. Lett.*, 2007, **17**, 3575.

130. W. Guba, W. Neidhart and M. Nettekoven, *Bioorg. Med. Chem. Lett.*, 2005, **15**, 1599.

131. L. S. Harikrishnan, M. G. Kamau, T. F. Herpin, G. C. Morton, Y. Liu, C. B. Cooper, M. E. Salvati, J. X. Qiao, T. C. Wang, L. P. Adam, D. S. Taylor, A. Y. A. Chen, X. Yin, R. Seethala, T. L. Peterson, D. S. Nirschl, A. V. Miller, C. A. Weigelt, K. K. Appiah, J. C. O'Connell and R. M. Lawrence, *Bioorg. Med. Chem. Lett.*, 2008, **18**, 2640.

132. D. C. Cole, J. R. Stock, R. Chopra, R. Cowling, J. W. Ellingboe, K. Y. Fan, B. L. Harrison, Y. Hu, S. Jacobsen, L. D. Jennings, G. Jin, P. A. Lohse, M. S. Malamas, E. S. Manas, W. J. Moore, M. M. O'Donnell, A. M. Olland, A. J. Robichaud, K. Svenson, J. Wu, E. Wagner and J. Bard, *Bioorg. Med. Chem. Lett.*, 2008, **18**, 1063.

133. T. D. Heightman, J. S. Scott, M. Longley, V. Bordas, D. K. Dean, R. Elliott, G. Hutley, J. Witherington, L. Abberley, B. Passingham, M. Berlanga, M. de los Fraires, A. Wise, B. Powney, A. Muir, F. McKay, S. Butler, K. Winborn, C. Gardner, J. Darton, C. Campbell and G. Sanger, *Bioorg. Med. Chem. Lett.*, 2007, **17**, 6584.

134. H. A. Vaccaro, Z. Zhao, J. W. Clader, L. Song, G. Terracina, L. Zhang and D. A. Pissarnitski, *J. Comb. Chem.*, 2008, **10**, 56.

135. C. A. Metcalf III, C. J. Eyermann, R. S. Bohacek, C. A. Haraldson, V. M. Varkhedkar, B. A. Lynch, C. Bartlett, S. M. Violette and T. K. Sawyer, *J. Comb. Chem.*, 2000, **2**, 305.

136. K. B. Sharpless and R. Manetsch, *Expert Opin. Drug Discovery*, 2006, **1**, 525.

137. R. Tietze, S. Lober, H. Hubner, P. Gmeiner, T. Kuwert and O. Prante, *Bioorg. Med. Chem. Lett.*, 2008, **18**, 983.

138. J. R. Morphy, Z. Rankovic and D. C. Rees, *Tetrahedron Lett.*, 1996, **37**, 3209.

139. T. Guo, Y. Shao, G. Qian, L. L. Rokosz, T. M. Stauffer, R. C. Hunter, S. D. Babu, H. Gu and D. W. Hobbs, *Bioorg. Med. Chem. Lett.*, 2005, **15**, 3696.

140. D. Maclean, F. Holden, A. M. Davis, R. A. Scheuerman, S. Yanofsky, C. P. Holmes, W. L. Fitch, K. Tsutsui, R. W. Barrett and M. A. Gallop, *J. Comb. Chem.*, 2004, **6**, 196.

141. C. A. Willoughby, S. M. Hutchins, K. G. Rosauer, M. J. Dhar, K. T. Chapman, G. G. Chicchi, S. Sadowski, D. H. Weinberg, S. Patel, L. Malkowitz, J. Di Salvo, S. G. Pacholok and K. Cheng, *Bioorg. Med. Chem. Lett.*, 2002, **12**, 93.

142. Y. Ambroise, B. Yaspan, M. H. Ginsberg and D. L. Boger, *Chem. Biol.*, 2002, **9**, 1219.

143. M. Congreve, G. Chessari, D. Tisi and A. J. Woodhead, *J. Med. Chem.*, 2008, **51**, 3661.
144. C. W. Murray and D. C. Rees, *Nat. Chem.*, 2009, **1**, 187.
145. P. T. Corbett, J. Leclaire, L. Vial, K. R. West, J.-L. Wietor, J. K. M. Sanders and S. Otto, *Chem. Rev.*, 2006, **106**, 3652.
146. M. S. Congreve, D. J. Davis, L. Devine, C. Granata, M. O'Reilly, P. G. Wyatt and H. Jhoti, *Angew. Chem., Int. Ed.*, 2003, **42**, 4479.
147. M. T. Cancilla, M. M. He, N. Viswanathan, R. L. Simmons, M. Taylor, A. D. Fung, K. Cao and D. A. Erlanson, *Bioorg. Med. Chem. Lett.*, 2008, **18**, 3978.
148. M. Hochgurtel, R. Biesinger, H. Kroth, D. Piecha, M. W. Hofmann, S. Krause, O. Schaaf, C. Nicolau and A. V. Eliseev, *J. Med. Chem.*, 2003, **46**, 356.
149. R. Frechette, *Annu. Rep. Med. Chem.*, 2007, **42**, 349.
150. Q. Ziao and D. Pei, *J. Med. Chem.*, 2007, **50**, 3132.
151. P. L. Richardson, *Curr. Pharm. Des.*, 2002, **8**, 2559.
152. J. Kofoed and J.-L. Reymond, *Chem. Commun.*, 2007, 4453.
153. J. Kurreck, *Angew. Chem., Int. Ed.*, 2009, **48**, 1378.
154. A. D. Miller, *Angew. Chem., Int. Ed.*, 1998, **37**, 1768.
155. A. Akinc, A. Zumbuehl, M. Goldberg, E. S. Leshchiner, V. Busini, N. Hossain, S. A. Bacallado, D. N. Nguyen, J. Fuller, R. Alvarez, A. Borodovsky, T. Borland, R. Constien, A. de Fougerolles, J. R. Dorkin, K. N. Jayaprakash, M. Jayaraman, M. John, V. Koteliansky, M. Manoharan, L. Nechev, J. Qin, T. Racie, D. Raitcheva, K. G. Rajeev, D. W. Y. Sah, J. Soutschek, I. Toudjarska, H.-P. Vornlocher, T. S. Zimmermann, R. Langer and D. G. Anderson, *Nat. Biotechnol.*, 2008, **26**, 561.
156. M. A. Snowden and D. V. S. Green, *Curr. Opin. Drug Discovery Dev.*, 2008, **11**, 553.
157. http://www.drugresearcher.com/Emerging-targets/GSK-opens-UK-s-largest-chemistry-lab.
158. S. Houlton, *Chem. World*, 2007, **4**(5), 52.
159. J. P. Nandy, M. Prakesch, S. Khadem, P. T. Reddy, U. Sharma and P. Arya, *Chem. Rev.*, 2009, **109**, 1999.
160. L. Weber, *Drug Discovery Today*, 2002, **7**, 143.
161. H. N. Weller, D. S. Nirschl, E. W. Petrillo, M. A. Poss, C. J. Andres, C. L. Cavallaro, M. M. Echols, K. A. Grant-Young, J. G. Houston, A. V. Miller and R. T. Swann, *J. Comb. Chem.*, 2006, **8**, 664.

CHAPTER 3

High Throughput Reaction Screening

ANDREW I. MORRELL

Worldwide Medicinal Chemistry, Pfizer Ltd., Ramsgate Road, Sandwich, CT13 9NJ, UK

3.1 Introduction

Synthetic chemistry plays a pivotal role throughout the lifecycle of a pharmaceutical product from the initial identification of a compound of interest to the large-scale manufacturing of an active pharmaceutical ingredient (API). Once a compound has been selected for progression through the drug development process towards becoming a marketed drug, the chemical structure and composition of that compound is unlikely to change although the synthetic route by which it is prepared almost certainly will. For example, although the chemical structure of Maraviroc™ **1** in the commercial HIV medicine marketed as Celsentri™ (Selzentry™ in the United States) is unchanged from the compound first synthesised in the research laboratory the original synthetic route has evolved significantly (Figure 3.1).[1]

Changes to the synthetic route as a compound makes progress towards becoming a marketed product are due to the commercial necessity to manufacture the finished API as cost-effectively, efficiently and safely as possible. In order to achieve this, the industrial chemist is required to evaluate reaction conditions and alternative synthetic routes in order to arrive at an optimised process for the preparation of the API. Therefore through the identification of optimal chemical processes the chemist can directly contribute to the reduction of compound attrition during the development phase of a pharmaceutical product.

RSC Drug Discovery Series No. 11
New Synthetic Technologies in Medicinal Chemistry
Edited by Elizabeth Farrant
© Royal Society of Chemistry 2012
Published by the Royal Society of Chemistry, www.rsc.org

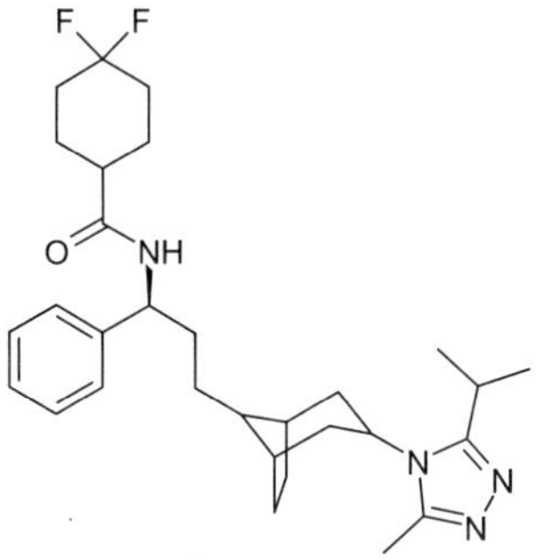

Figure 3.1 The structure of Pfizer's Maraviroc™.

With the advent of high throughput chemistry and biology in the pharmaceutical industry over the last decade or so, organic chemists have a wide range of tools and techniques available to them to carry out chemical reactions more quickly and on a smaller scale than ever before. One notable aspect of these developments has been the widespread adoption of parallel chemistry techniques, whereby large numbers—occasionally thousands—of discrete chemical processes are carried out simultaneously to prepare individual compounds for biological evaluation. In conjunction with this drive towards greater throughput of chemical synthesis, the closely associated analytical and purification technologies have also evolved to meet the demand for faster analysis, isolation and characterisation of reaction products.

Whilst many of these high throughput chemistry technologies were initially developed for the synthesis and purification of large numbers of analogous compounds, typically referred to as *compound libraries*, more recently the same techniques have found an application in the optimisation of individual compound syntheses. A significant advantage of adopting these technologies is that through the miniaturisation of each individual experiment, the overall consumption of material during the optimisation process is minimised. This is of particular importance in an exploratory chemistry setting where material supply may be limited. Reducing the quantity of material consumed during process optimisation also reduces the quantity of waste produced which has both economic and environmental benefits.

For a given chemical process we can assume that the starting materials and the desired product will remain constant unless there is a fundamental redesign of the synthetic route. In order to optimise the process, the chemist can examine the effect of altering other variables such as reagent, solvent and temperature upon the outcome of the process. By taking a high throughput, parallel approach to optimisation the chemist can evaluate, or *screen*, a large number of chemical process variables simultaneously. This initial screening can identify conditions suitable for further investigation and through additional experimentation a set of optimum conditions for that process can be derived.

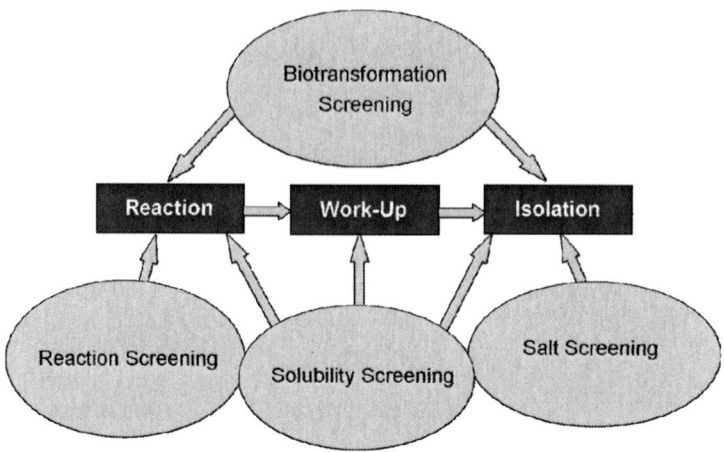

Figure 3.2   Opportunities for screening in a typical synthetic process.

A typical synthetic process can be considered to consist of three distinct practical stages:

1. The 'reaction' stage: the transformation of starting material into product, usually by the action of a reagent upon that material.
2. The 'work-up' stage: cessation of the chemical process and may include a preliminary separation of the desired product from other reaction components, for example by liquid-liquid extraction.
3. The 'isolation' stage: isolation and purification of the desired product and may involve chromatographic or non-chromatographic separation processes.

Each of these stages offers multiple opportunities to optimise the process using a variety of screening strategies as shown in Figure 3.2. An improvement at any of these stages would be expected to lead to an overall improvement in that process. A discussion of the main approaches to arriving at an optimised chemical process using high throughput screening follows.

3.2   High Throughput Reaction Screening

For the synthetic organic chemist, reaction screening is probably the most obvious application of a high throughput experimental approach for any chemical process optimisation. The optimisation of a chemical reaction would typically involve a study of the effects of different reagents and solvents on the outcome of the process. The outcome of the process or *response* typically of most interest to the synthetic chemist is the yield of the desired product.

Reaction process variables such as individual reagents or solvents are referred to as 'discontinuous' or 'discrete' variables or *factors*. Variables such as reagent stoichiometry, reaction temperature and concentration, which can be selected from a range of possible settings within a process, are known as 'continuous' variables or factors. A combination of all of possible variable factors for a chemical reaction at all levels is often referred to as the 'reaction space', which is defined by the upper and lower settings of the variables.

The conventional approach to identifying an optimum process would involve selecting a combination of reagents and solvents that was predicted to deliver the highest yield of the desired reaction product. If an acceptable result was not obtained further experiments would be carried out, typically changing one factor at a time, and this activity would be repeated until the desired result was obtained. Depending upon the time and quantity of material available, the chemist may identify the optimum set of reaction conditions from within the set of experiments completed. One limitation with this serialised approach to reaction optimisation is that it may not be readily apparent when an 'optimum' has been reached, or indeed when to cease further experimentation.

The simplest type of reaction screen would be one where a single-variable process parameter is examined. As an example, for the base-promoted Williamson ether synthesis of anisole **4** from phenol **2** and iodomethane **3** illustrated in Scheme 3.1, one could begin with the hypothesis that the base was the most important factor in determining the outcome of this reaction.

In this example a set of selected bases could be assessed for their effectiveness in forming product **4** and all other process variables could be held constant. The optimum base from within the screening set could be identified in this manner, although as all other variables were held constant the *overall* process may not be optimal. A more powerful approach would be to vary several process parameters simultaneously and this is where high throughput, parallel screening offers real advantages over 'one variable at a time' optimisation.

An efficient method of reaction screening and optimisation would be to preselect a set of reagents and solvents and design a discrete set of experiments to examine different combinations of these in conjunction with other process variables, such as temperature. The experimental parameters could be selected using either the existing knowledge of a process or to test a specific hypothesis. From within this defined set of experiments it would be possible to select the optimum conditions which could be applied directly to the process of interest, or further refined by additional experimentation.

The application of high throughput reaction screening is of particular benefit in optimising complex multi-parameter processes, and this is reflected in the

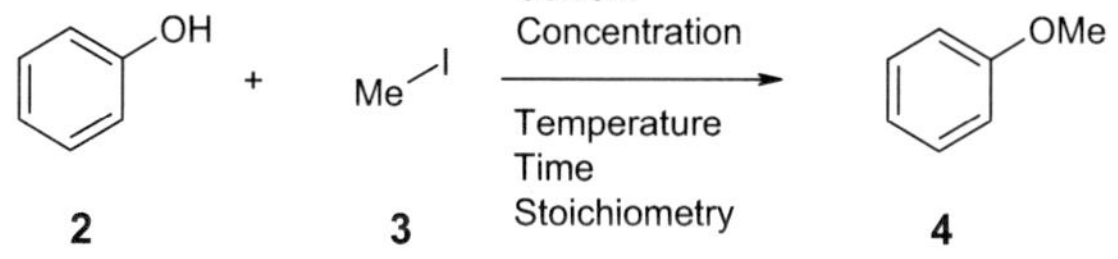

Scheme 3.1 Williamson ether synthesis – variable process parameters.

fact that a number of published examples in this field relate to the optimisation of transition-metal-mediated cross-couplings where the number of possible process variables can be considerable.

3.2.1 Statistical Design of Experiments in Reaction Screening

The use of statistical design of experiments (DoE) is an area that is growing in popularity and importance within the synthetic chemistry community.[2] The aim of DoE is to determine how each of the process variables (factors) affects the desired outcome, or *response*, from a process and generally to optimise that response whilst carrying out the minimum number of experiments.

The DoE approach to optimisation can be used to examine several factors at different settings, for example looking at the effect of low and high temperature upon a chemical reaction. A two-level, three-factor (*i.e.* three variables) central composite face (CCF) design typical of those used in reaction screening is shown in Figure 3.3.

Each axis in this design represents a separate factor, and each sphere represents a combination of the factors being explored at different settings as a single experiment. For a chemical reaction, the volume within the cube represents the 'reaction space'. The experiment represented by the centre-point of the cube is usually replicated to determine process reproducibility. Depending upon the complexity of the design, additional points (*i.e.* experiments) can be added to examine intermediate settings for each factor and how they affect the response.

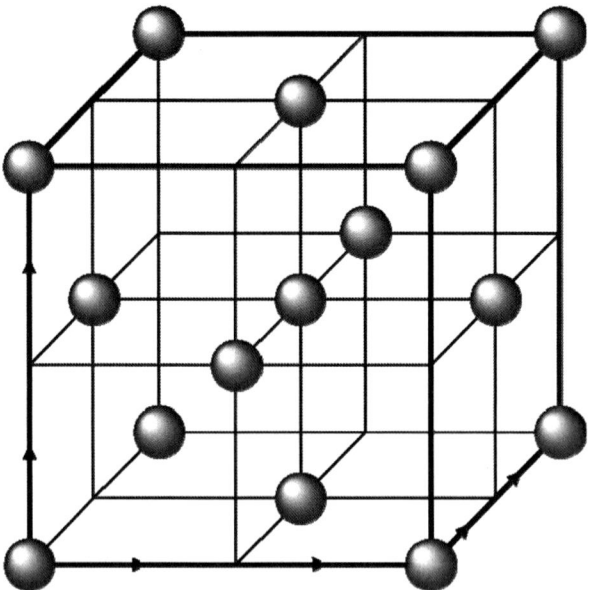

Figure 3.3   A two-level, three-factor central composite face (CCF) experiment design.

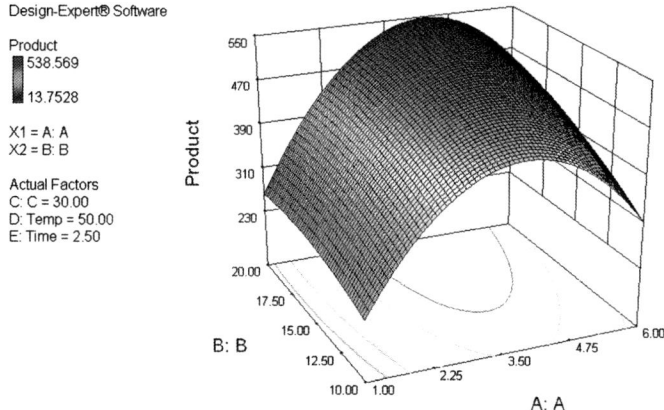

Figure 3.4 An example of a three-dimensional response surface plot showing the variation of product response with factors A and B.

Once the experiments have been carried out the results can be processed using DoE software and plotted as a three-dimensional 'response surface' as shown in Figure 3.4. The surface is fitted to the known data-points and by visualising the data in this way trends within the process can be identified. It is also possible to use DoE predictively, within the reaction space examined, to identify specific combinations of factor settings which would be optimal for a maximum response. Perhaps the most attractive aspect of DoE to the synthetic chemist is this potential for predicting the 'optimum' set of conditions from a limited number of experiments.

DoE was used by workers at Argonaut to study and optimise a solid-phase amidation reaction for use in compound library synthesis.[3] In the process of optimisation a solid-phase reagent **5** was acylated with a carboxylic acid **6** to provide an 'activated ester' **7** (Scheme 3.2). Addition of an amine **8** to this activated ester intermediate led to the formation of a carboxamide **9**.

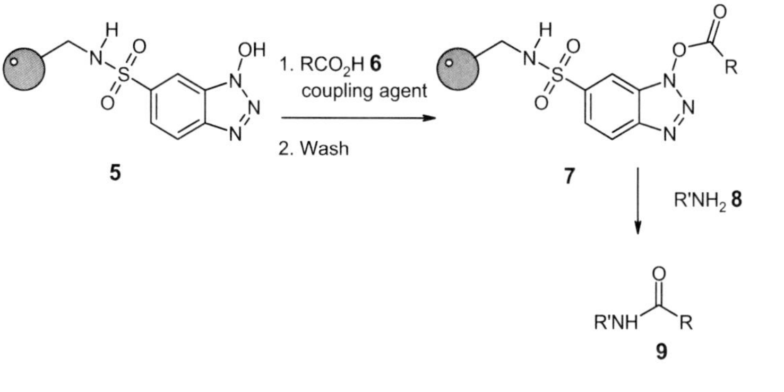

Scheme 3.2 Solid-phase amidation process optimised using DoE.

By carrying out several sets of statistically designed experiments the coupling agent, carboxylic acid stoichiometry, order of reagent addition and reaction time were all examined. The data from the experiments were analysed using DoE software to generate predictive three-dimensional response surface models and identify the optimum settings for all of these factors. The optimum conditions were then successfully applied to the parallel synthesis of a large series of amides.

3.2.2   Suzuki–Miyaura Reaction Catalyst Screening

The Suzuki–Miyaura reaction is one of the most widely used methods for aromatic carbon–carbon bond formation in modern synthesis and the number of catalytic systems available is continually increasing.[4,5] A simple example of the Suzuki–Miyaura reaction in which phenylboronic acid **10** is cross-coupled with bromobenzene **11** to form biphenyl **12** is illustrated in Scheme 3.3.

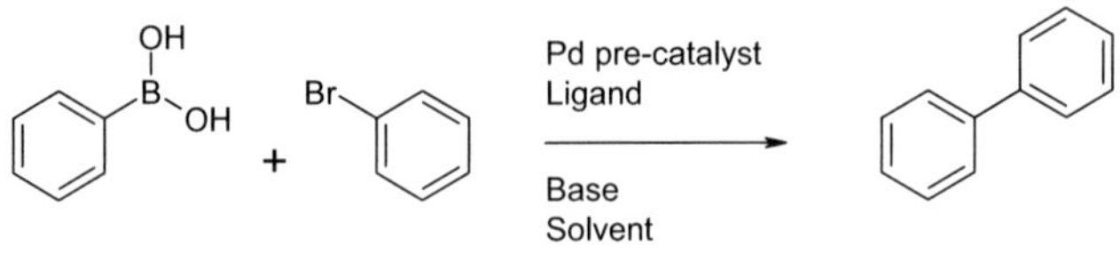

Scheme 3.3   A simple Suzuki–Miyaura cross-coupling.

Since the reaction process typically involves two coupling partners, a palladium salt 'pre-catalyst', ligand, base and solvent, it can be appreciated that the number of possible process variations is huge. Thus, the selection of an optimal set of reaction conditions from the literature related to the Suzuki–Miyaura reaction would be a lengthy task taking a stepwise optimisation approach. The Suzuki–Miyaura reaction is therefore an ideal candidate for high throughput reaction screening.

Process development chemists at Merck wanted to identify a high-yielding and regioselective process for the preparation of the 1,6-naphthyridone **15** using Suzuki–Miyaura cross-coupling methodology (Scheme 3.4).[6] The existing process involved the use of $Pd(OAc)_2$ and PPh_3 to form $Pd(PPh_3)_4$ as the catalyst, with K_3PO_4 as the base and isopropanol as the solvent. Although these conditions led to the desired product **15** a significant amount of the bisadduct **17** was also formed during the reaction. The overall conversion of **13** and **14** into the mixture of **15** and **17** was 93%, although since **15** and **17** were formed in a 2 : 1 ratio there was clearly scope for further optimisation of this process. Adjustments made to the existing process did not result in the desired product **15** being formed both selectively and in high yield.

A systematic high throughput screen of approximately 80 different catalyst systems was undertaken and included all of the main ligand classes with

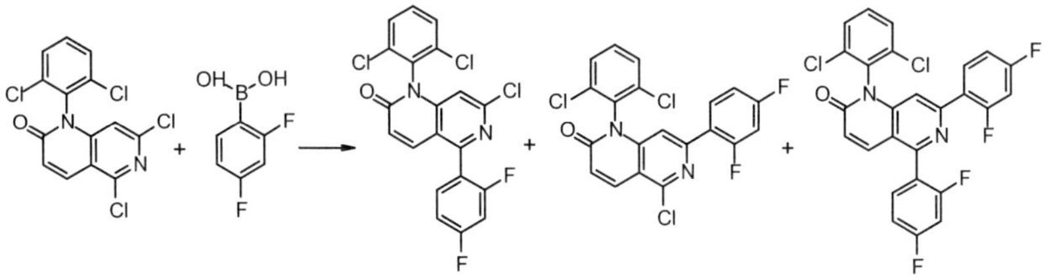

Scheme 3.4 Suzuki–Miyaura cross-coupling between **13** and **14**.

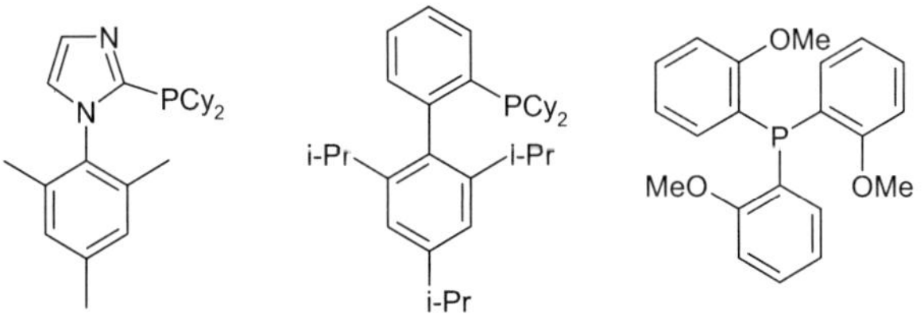

Figure 3.5 Selected structures of ligands used in the Suzuki–Miyaura optimisation study.

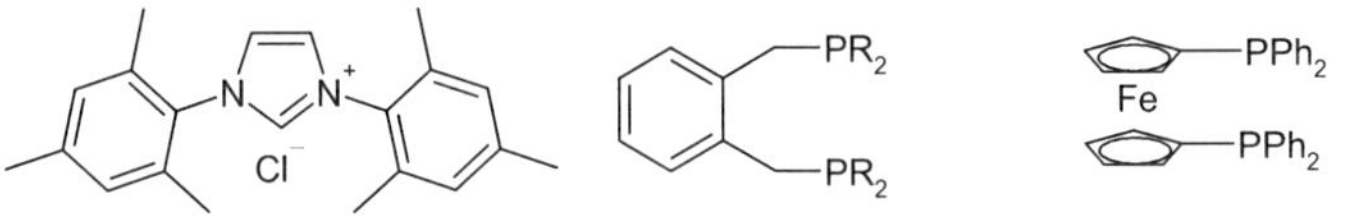

precedent in the Suzuki–Miyaura reaction. Within this set were groups of analogous ligands which would explore steric and electronic effects upon the product profile and a selection of these ligands is illustrated in Figure 3.5.

The authors note that for many projects at an early stage in the drug development process, the amount of material available for this type of study is often limited. After selecting a set of ligands the number of possible process variations was still considerable and therefore the authors intentionally selected a limited range of four bases and four solvents from those commonly reported in the literature. In order to further limit the number of possible experiments, two temperature settings and a single 'catalyst loading' of 2.0 mol% were selected.

From the results presented it was apparent that within the set of reactions screened it was possible to obtain high conversion of the starting materials **13** and **14** into the coupled products **15**, **16** and **17**. Several catalyst systems gave a high degree of selectivity for one of the three possible products over the others; however, the steric and electronic effects of the ligands were subtle and would have been difficult to predict without carrying out this empirical study. The two structurally diverse ligands (2-MeO-Ph)₃P **20** and IMes **21** were selected for further study.

Additional experiments to optimise the process using ligands **20** and **21** were carried out, and again the complexity of the interactions between the various reaction parameters was revealed. It was found that both the solvent and the base had significant effects upon the conversion and regioselectivity of the cross-coupling using both ligand **20** and ligand **21**. Additionally it was found that the palladium *pre-catalyst* played an important role in the process, as did the relative proportions of ligand and palladium.

Due to the complexity of the interactions between each of the reaction variables revealed by screening, the authors then chose to study the process using a statistical DoE approach. At this point it is worth noting that ligand **20** was selected for further study in preference to ligand **21** due to its lower relative cost and higher catalytic turnover compared to **20**. These two considerations make it a more attractive ligand for use in an industrial-scale process.

Following this DoE study the authors were able to identify a set of conditions which gave the desired reaction and product profile (Figure 3.6). It can be appreciated from this example that an initial broad-based screen gave valuable

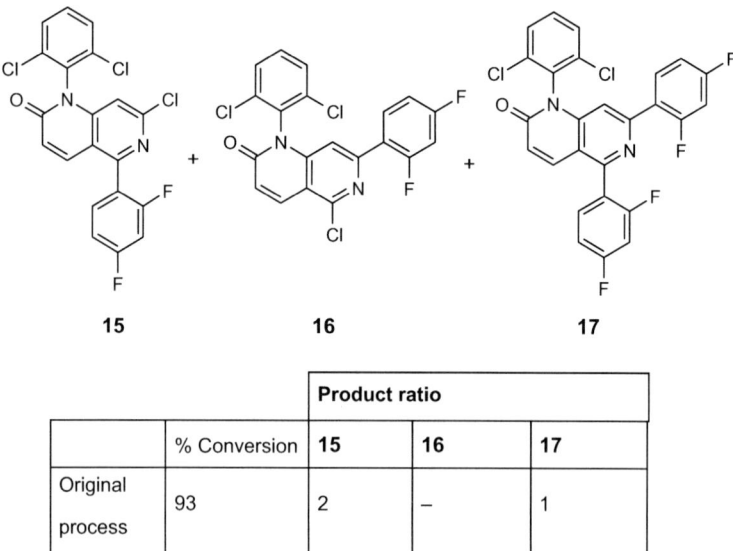

		Product ratio		
	% Conversion	15	16	17
Original process	93	2	–	1
Optimised process	98	92	3	5

Figure 3.6 Products obtained before and after optimisation.

insights into alternative catalyst systems, which could be subsequently optimised using a focused screening approach in combination with statistical experimental design methods.

3.2.3 Heck Reaction Catalyst Screening

A similar approach to the optimisation of a Heck reaction was taken by Hartwig and co-workers.[7] In this example high throughput screening was used to identify catalysts with potential for carrying out the Heck reaction at room temperature (Scheme 3.5).

The 'dansyl' and azo-dye groups fulfil the roles of fluorophore and quencher in **24** and **25** respectively to facilitate the fluorescence resonance energy transfer (FRET) detection method used to rapidly quantify the product formed (**26**), and hence determine the catalyst activity during this study.

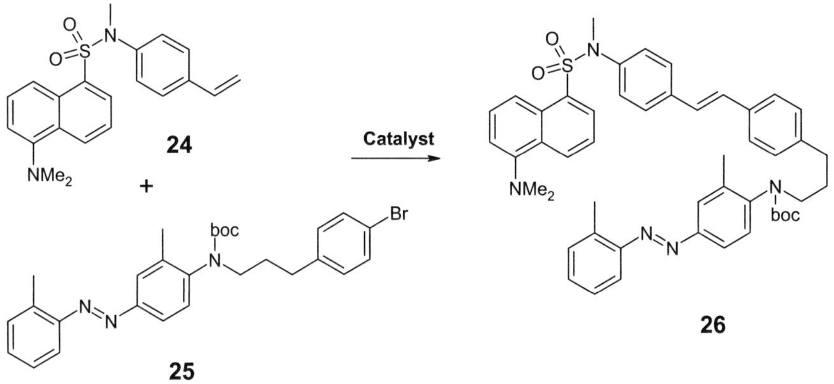

Scheme 3.5 Heck reaction studied using catalyst screening.

A set of 96 Heck reactions were assembled and carried out in parallel in a 96-well glass plate. Each well contained a different phosphine ligand and all other reaction variables were kept constant. The glass reaction plate was analysed using an automated fluorescent plate reader and, since the product shows weak fluorescence compared to **24**, this method allowed the rapid quantitative assessment of the effectiveness of each of the 96 phosphines screened. The most effective ligands are shown in Figure 3.7.

The initial screen highlighted the activity of both ferrocenyl- (**28** and **29**) and adamantyl-substituted phosphines in this process and therefore 1-adamantyl-*tert*-butyl ferrocenyl phosphine **33** was synthesised for inclusion and evaluation in further experiments.

A total of 16 ligands were selected for additional screening in 6 different solvents to examine the effect of the solvent on the outcome of each reaction. Interestingly the solvent appeared to exert minimal influence over the process. Two of the most 'active' ligands to emerge from these two screens, **28** and **31**,

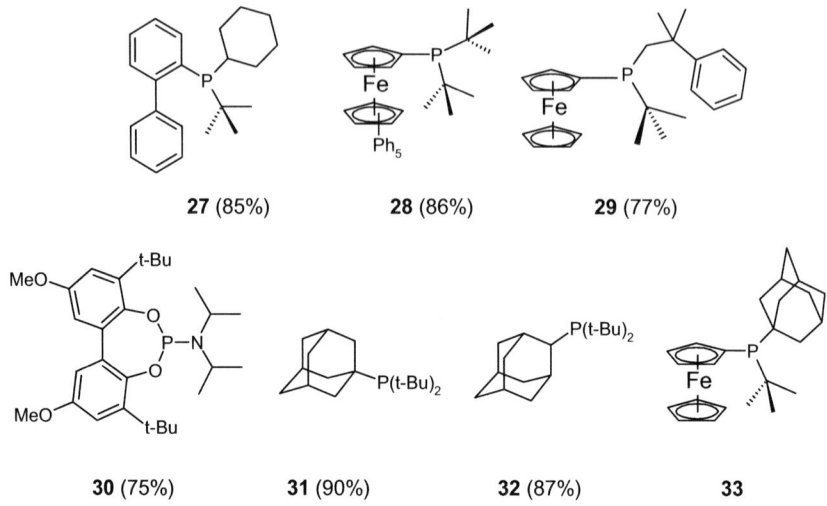

27 (85%)   **28** (86%)   **29** (77%)

30 (75%)   **31** (90%)   **32** (87%)   **33**

Figure 3.7 Examples of ligands giving high yields of product **14** (% yields in parentheses).

were then assessed at room temperature in a range of Heck reactions and found to be effective catalysts for a range of substrates. Thus through a systematic series of screens the authors were able to identify the optimal catalysts and conditions for this important transformation.

3.2.4 Hydrogenation Catalyst Screening

Another application of high throughput reaction screening to identify an optimised catalytic process was reported by Hawkins and Makowski.[8] In this example a two-stage parallel screen of hydrogenation catalysts and conditions was carried out to identify optimum processes for the selective conversion of 4-nitroacetophenone **34** into each of the three possible reduction products **35**, **36** and **37** (Scheme 3.6).

A set of eight different hydrogenation catalysts were selected for evaluation and these catalysts were screened both in the presence and absence of methanesulfonic acid (MsOH) at a single catalyst loading, temperature, hydrogen pressure and reaction concentration in order to simplify the experiments.

The experiments were carried out in a commercially available parallel pressure reactor. From an initial screening it was apparent that significantly different proportions of the three possible products **35**, **36** and **37** could be obtained, depending upon the catalyst and additive combination used (Table 3.1).

Three different sets of conditions selected from this initial screen were examined in a second round of experiments to optimise the selective production of each of the three possible products **35, 36** and **37**. By examining high and low

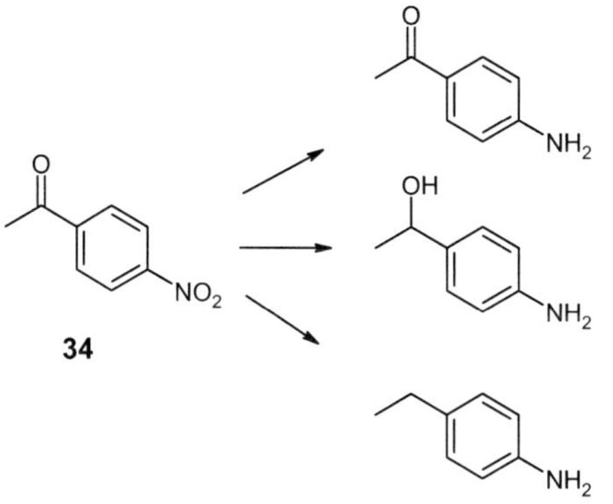

Scheme 3.6 Hydrogenation products from 4-nitroacetophenone.

Table 3.1 Catalyst screen results for the hydrogenation of 4-nitroacetophenone **34**.

		In situ *yield*[b]		
Catalyst[a]	Equiv. MsOH	25	26	27
5% Pd/C	0	14	54	21
	1.1	19	21	57
10% Pd/C	0	7	**90**	0
	1.1	0	2	**96**
20% Pd(OH)$_2$/C	0	8	64	16
	1.1	1	1	86
5% Pd/CaCO$_3$/Pb	0	**85**	7	1
	1.1	66	4	26
5% Pt/C	0	66	2	2
	1.1	6	37	53
PtO$_2$	0	70	20	1
	1.1	4	7	81
5% Rh/C	0	54	39	2
	1.1	25	21	52
5% Rh/Al$_2$CO$_3$	0	59	34	2
	1.1	19	31	48

[a]10% loading, 30 °C, 50 psi, EtOH, 22 h, 500 mg scale.
[b]Determined by HPLC with reference to an external standard.

settings for the continuous variables of catalyst loading, hydrogen pressure, temperature and MsOH acid level for each of the three sets of conditions using a DoE approach it was possible to obtain **35, 36** and **37** in 97%, 95% and 99% yield respectively.

This is a good example of the use of a multi-stage, high throughput screening approach firstly to identify a discontinuous process variable—in this case the catalyst—and secondly to optimise the process by adjusting the continuous variables within defined limits based upon an understanding of the process gained through initial broad-based screening.

3.2.5   Biotransformation Reaction Screening

Although the use of enzymes in organic chemistry reactions, commonly referred to as 'biotransformations', has been an established technique for over one
hundred years, there has been an enormous increase in interest in this area in
recent years and there are now numerous examples of their use in large-scale
preparations.[9] Biotransformations are typically catalytic, highly chemo- and
stereoselective processes and are usually carried out in an aqueous environment. All of these factors combine to make a strong case for incorporating a
biotransformation step into a synthetic route where possible, particularly from
an environmental perspective.

Scientists at Pfizer have demonstrated that the automated high throughput
screening of enzymes can successfully be used to identify and optimise biotransformation processes.[10] Using automated liquid handling instruments, sets
of enzymes can be dispensed as solutions into 96-well plates to provide enzyme
screening 'kits'.

For increased efficiency, large numbers of these screening 'kits' can be prepared in batches, and once prepared these plates can be conveniently stored
below –20 °C for extended periods with minimal deterioration. For example,
the viability of a lipase screening kit was studied and it was found that, for the
majority of enzymes, over 90% of the original activity was retained after a
single freeze–thaw cycle. To carry out the screening process the substrate is
simply added as a stock solution to a thawed screening plate after the addition
of an appropriate buffer. The plate is then incubated at 30 °C and aliquots can
be periodically removed for analysis.

This approach to high throughput biotransformation screening and optimisation is exemplified by a kinetic resolution introduced into the synthetic
route to the pharmaceutical product Pregabalin™ **41**, the active ingredient in
Pfizer's Lyrica™ by Pfizer scientists.[11] The proposed new route to **41** involved
the introduction of an enzymatic kinetic resolution step for the racemic synthetic intermediate **38**. Pregabalin™ had previously been obtained as a single
enantiomer using a classical salt resolution of racemic **39** with (*S*)-(+)-mandelic
acid during the 'first-generation' synthetic route illustrated in Scheme 3.7.

A set of commercially available hydrolase enzymes was screened in a 96-well
plate format in order to identify an enzyme that would offer a kinetic resolution of
racemic **38** into *R*-**38** and **40**. The enzymes were all screened at a substrate loading
of 5% v/v and at an initial pH of 7.2. A number of enzymes in the screening plate
were effective in hydrolysing **38**, although only a limited number offered the
required high degree of enantioselectivity. The lipase from *Thermomyces lanugi-
nosus*, which is available commercially as Lipolase™, offered the highest degree of
enantioselectivity and enzymatic activity from the enzymes screened.

Once a suitable enzyme had been identified by screening, it was necessary to
further optimise the process in terms of the other important process variables
such as enzyme loading, concentration, pH and solvent composition.

Once optimised, the process was tested for robustness, or reproducibility, by
repeating it over twenty separate runs each at the 10 kg scale of **38**, before

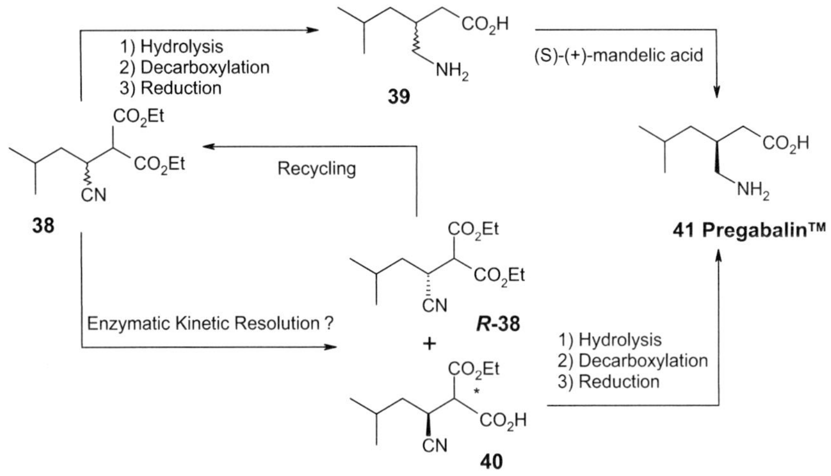

Scheme 3.7 The 'first-generation' route and proposed enzymatic route to Pregabalin™.

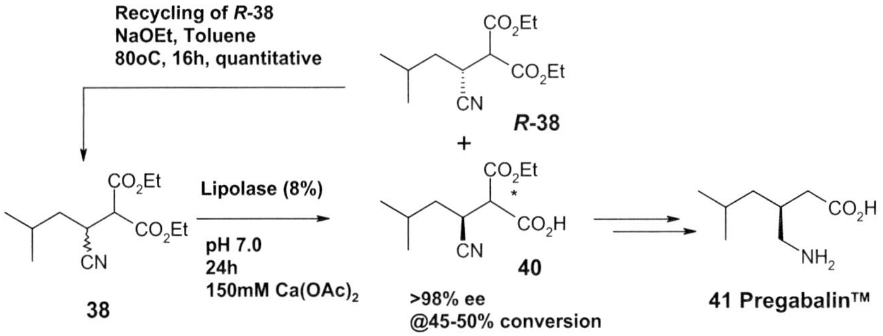

Scheme 3.8 Optimised Pregabalin™ synthesis incorporating a biotransformation step.

further scale-up to 900 kg and finally an impressive 3.5 tonne scale. The optimised process included an efficient recycling step for the undesired isomer *R*-**38** (Scheme 3.8).

Overall the introduction of the biotransformation step into the new route to Pregabalin™ allows for a more efficient process since only the desired isomer of **38** is carried through the final steps of the synthesis. This gives a significant reduction in the quantity of reagents used in the latter steps of the synthesis and consequently the amount of waste material produced from the new route.

Thus, in a similar manner to that described for conventional chemical reaction screening, a potential biotransformation of interest can be screened against a number of individual enzymes. Those enzymes yielding the desired result can be studied further and the biotransformation process can be optimised.

3.2.6 Reaction Screening Using Continuous Processes

There has been significant recent interest in the use of both continuous-flow (CF) and plug-flow (PF) reactor technology in organic synthesis, and its application to reaction optimisation.[12,13] These 'flow-chemistry' technologies typically involve carrying out chemical processes within small diameter tubing, where the reaction mixture is mobilised and may be subjected to temperature or pressure variations during an experimental run. Flow reactors offer significant potential advantages over conventional discrete reactor technologies, often referred to as 'batch' processes, including precise parameter control and extended parameter limits. These advantages are ideal for reaction screening as process parameters can be continuously varied during a single optimisation experiment.[12]

An example of the use of a continuous-flow reactor to rapidly screen and optimise the synthesis of a series 1,4-disubstituted 1,2,3-triazoles was published recently by Bogdan and Sach.[14] The copper-catalysed cycloaddition of an organic azide with an acetylene is employed for the preparation of 1,4-disubstituted 1,2,3-triazoles. The conventional 'batch' process would typically involve the mixing of an organic halide **42** and sodium azide to form an organic azide **43**, which would be added to an acetylene **44** in the presence of a copper source to form a 1,4-disubstituted 1,2,3-triazole **45** (Scheme 3.9).

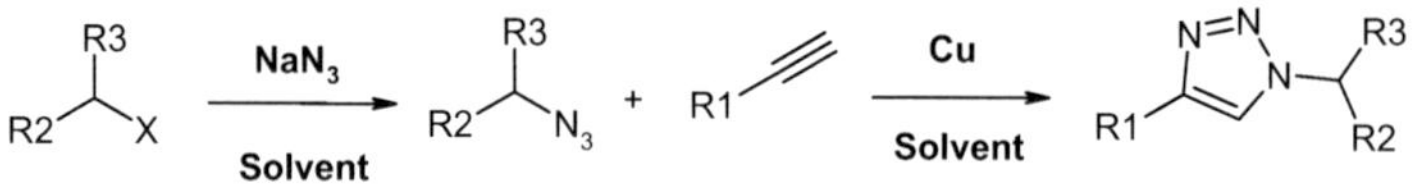

Scheme 3.9 Copper-catalysed triazole formation.

One disadvantage of using this process is that organic azides are a potentially hazardous class of reagent. By carrying out the formation of the organic azide *in situ* within a continuous-flow reactor in the presence of the acetylene component, a potentially hazardous accumulation of the azide would be avoided, since only a small quantity would be generated immediately prior to the cycloaddition.

The reaction selected for optimisation using continuous flow parameter screening is shown in Scheme 3.10. A copper-coil flow reactor was custom-built as previous studies had indicated that this reaction could be catalysed by copper metal. After passing a mixture of **46**, **47** and sodium azide as a solution in *N,N*-dimethylformamide (DMF) through the heated copper-coil reactor, it was found that the internal surface of the copper tubing was indeed able to catalyse the desired cyclisation to form **48**. By screening of the process variables for this reaction and applying DoE analysis, an optimum process which produced **48** in 71% yield was rapidly identified.

The introduction of microwave-assisted organic synthesis (MAOS) into the chemistry laboratory has allowed the organic chemist to perform a number of

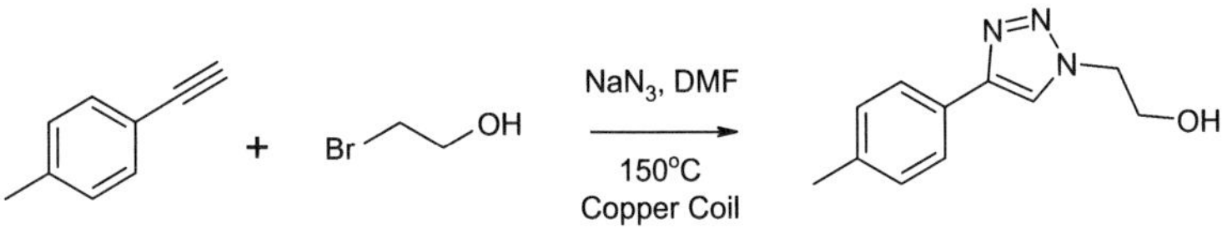

Scheme 3.10   The triazole formation optimised using a copper-coil flow reactor.

synthetic transformations rapidly and under conditions that otherwise would be difficult to achieve.[15] However, one possible limitation has been the difficulty in effectively scaling-up microwave-optimised conditions for preparative synthesis.

A relatively recent development in this area has been the 'stop-flow' microwave reactor, which is amenable to both rapid reaction screening and subsequent scale-up, and interesting examples of the use of fluorous 'spacer' methodology for reaction optimisation in a plug-flow microwave reactor were recently published by scientists at GlaxoSmithKline (GSK).[13,16] Several chemical processes were optimised by passing discrete reaction 'plugs' in an organic solvent separated by an immiscible perfluorinated solvent through a microwave reactor. The plug-flow method has the advantage over continuous-flow reaction screening and optimisation that very small scale reactions can be carried out during the screening phase, saving potentially valuable material. Once the optimum process has been developed, scale-up for preparative purposes can be achieved simply by increasing the volume of the reaction plug.

The use of flow reactors in reaction screening, optimisation and scale-up is an area that is likely to develop further as this technology gains wider acceptance by the synthetic chemistry community.

3.3   High Throughput Salt Screening

The formation of a salt of an ionisable compound (salification) typically alters its physicochemical properties and this can be exploited for the purification of a compound by crystallisation. This principle can be applied to the separation of ionisable racemic compounds by using an appropriate single-enantiomer counter ion. Crystallisation and filtration is an attractive purification method in large-scale chemical processing due to throughput and cost constraints associated with chromatographic purification methods.

The identification of a salt form of an ionisable pharmaceutical product may enhance important properties such as aqueous solubility and bioavailability, simplify manufacturing and also expand the intellectual property domain of that product. Therefore salt-form identification can be one of the most important activities during the drug development process.

For all of these applications, counter-ion selection from the vast number of available acids and bases would be facilitated by adopting a high throughput

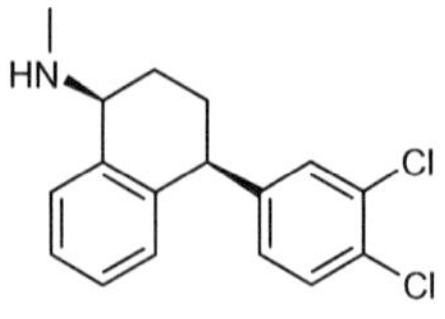

49

Figure 3.8 Sertraline (Zoloft).

Table 3.2 Components used in high-throughput salt screening of Sertraline.

Solvents	Monoprotic acids	Polyprotic acids
Acetonitrile	Acetic acid	Citric acid
Ethanol	Benzenesulphonic acid	fumaric acid
iso-propanol	Benzoic acid	Maleic acid
iso-butanol	Ethanesulphonic acid	Malonic acid
iso-propyl acetate	Hydrobromic acid	Phosphoric acid
Tetrahydrofuran	Methanesulfonic acid	Succinic acid
n-heptane	Lactic acid	Sulphuric acid
Propylene glycol	*p*-toluenesulphonic acid	L-tartaric acid
Water		

screening approach. The fundamental approach for salt selection is the same irrespective of the intended application: a selection of counter ions is screened for their capacity to form a crystalline salt from a solution of the ionisable 'parent' compound. These screens can be carried out in a multi-well plate format, and can be readily automated.[17]

A detailed study of automated high throughput salt screening for the identification of novel salt forms for Pfizer's Sertraline (Zoloft) **49** (Figure 3.8) has been reported.[18] Over 3600 automated crystallisation experiments were carried out using monoprotic and polyprotic pharmaceutically acceptable acidic counter-ions in a range of solvents (Table 3.2).

Using high throughput screening of these acids, 18 crystalline salt forms of Sertraline were identified, and information about their propensity for polymorphism was gained simultaneously. This study clearly demonstrates the value of high throughput salt screening for solid-form identification.

3.4 High Throughput Solubility Screening

High throughput screening of different solvents is an efficient way to identify the optimum solvent for a given process. This type of screen can identify optimal solvents for crystallisation, for example when optimising a crystallisation process during salt screening.

The relative solubility of different components in a solvent can also be exploited in order to greatly simplify the work-up and isolation stages of a process. The identification of a solvent that is good at dissolving starting materials but a poor solvent for the product of a process could be valuable, for example to precipitate the product as it is formed, perhaps driving a reversible reaction to completion and simplifying product isolation through a filtration process. This is known as a 'direct-drop' or 'direct-isolation' process.[19]

The procedure for carrying out solvent screening is straightforward and simply involves determining the relative solubility of the components of a process in a range of solvents. Solvents can be screened in parallel, and solvent dispensing can be readily automated.

3.5 Equipment and Automation Used in High Throughput Reaction Screening

The vessels used during screening and optimisation processes can range in volume from several microlitres to several millilitres depending upon the scale of the experiments. In our laboratories we have found that borosilicate glass vials with septa-equipped aluminium crimp-caps and individual magnetically-driven stirrer-bars provide suitable vessels for the vast majority of chemistry screens (Figure 3.9). Sets of these vials can be placed into metallic racks and, when used in conjunction with a stirrer–hotplate, they adequately resemble the ubiquitous round-bottomed flask encountered in the typical synthetic chemistry laboratory.

The crimp-cap provides a seal to contain the reaction components and the integral septum allows aliquots to be taken from the vessel whilst maintaining its integrity. These vials are commercially available in a range of volumes, allowing the scientist to select the most appropriate size for the intended application, and they are relatively inexpensive. They are compatible with most commonly-encountered well-plate formats and integrate easily with automated equipment.

The use of automation as a means to maximise the efficiency of chemistry techniques is well established within chemical research laboratories, and one of the first examples of the use of automation specifically for reaction screening and optimisation was published over a decade ago.[20] Consequently this area is

Figure 3.9 Vessels used for high-throughput chemistry screening.

well served by suppliers of instrumentation and consumables, and the chemistry-screening scientist has many options to choose from.

In our own laboratories we generally use individual automated instruments for each specific task within the screening and optimisation workflow. Our belief is that individual or modular automation systems comprising totally separate workstations for each task allow the incorporation of 'best-in-class' equipment at each stage of an optimisation process and allow more flexibility. In contrast, a single multi-functional instrument may involve compromises in its design which may limit its overall performance and reliability.

3.6 Analytical Techniques Used in High Throughput Reaction Screening

High-performance liquid chromatography (HPLC) is particularly suited to high throughput chemistry screening for a number of reasons.[21] Firstly it is rapid; using recently introduced fast LC methods and instruments it is possible to get high-quality data with short sample cycle-times. Secondly it is quantitative; comparison of the individual experiments with external standards allows accurate *in situ* quantitation and cross-comparison of results. Finally 'direct sampling' from vessels is possible using glass vials with septa crimp-caps specifically designed for HPLC instruments. Aliquots can be taken directly from a reaction vessel, and using a modified HPLC instrument such as that currently used in our own laboratories (Figure 3.10) it is possible to obtain accurate real-time profiling of a process. By coupling a HPLC instrument to a liquid chromatography mass spectrometer (LC-MS) it is also possible to obtain structural information and characterise unidentified components of a mixture.

Other analytical techniques that have found use in high throughput reaction screening experiments include thin-layer chromatography[22] and fluorescence resonance energy transfer (FRET).[6]

Figure 3.10   HPLC instrument modified for chemistry screening purposes.

3.7 Conclusion

The use of high throughput chemistry screening allows the rapid optimisation of many aspects of a chemical process. Initial broad-based screening can be used to identify sub-optimal processes or 'hits', and these processes can subsequently be fully optimised, for example through the integration of statistical design methods into screening. As a method for identifying optimised chemical processes it is efficient and requires little in the way of specialised laboratory equipment. By following this approach it is possible to fully optimise a chemical process with relatively few small-scale experiments which is very attractive from a time and economic perspective.

References

1. S. J. Haycock-Lewandowski, A. Wilder and J. Ahman, *Org. Process Res. Dev.*, 2008, **12**, 1094.
2. H. Tye, *Drug Discov. Today*, 2004, **9**, 485.
3. O. W. Gooding, L. Vo, S. Bhattacharyya and J. W. Labadie, *J. Comb. Chem.*, 2002, **4**, 576.
4. (a) N. Miyaura, K. Yamada and A. Suzuki, *Tetrahedron Lett.*, 1979, **36**, 3437; (b) A. Suzuki, T. Yanagi and N. Miyaura, *Synth. Commun.*, 1981, **11**, 513.
5. R. Martin and S. L. Buchwald, *Acc. Chem. Res.*, 2008, **41**, 1461.
6. C. Cai, J. Y. L. Chung, J. C. McWilliams, Y. Sun, C. S. Shultz and M. Palucki, *Org. Process Res. Dev.*, 2007, **11**, 328.
7. J. P. Stambuli, S. R. Stauffer, K. H. Shaughnessy and J. F. Hartwig, *J. Am. Chem. Soc.*, 2001, **123**, 2677.
8. J. M. Hawkins and T. W. Makowski, *Org. Process Res. Dev.*, 2001, **5**, 328.
9. (a) K. Drauz and H. Waldmann, *Enzyme Catalysis in Organic Synthesis*, Wiley, Weinheim, 2002; (b) A. Liese, K. Seelbach and C. Wandrey, *Industrial Biotransformations*, Wiley, Weinheim, 2nd edn., 2006; (c) U. T. Bornscheuer and R. J. Kazlauskas, *Hydrolases in Organic Synthesis*, Wiley, Weinheim, 2nd edn, 2006.
10. D. R. Yazbeck, J. Tao, C. A. Martinez, B. J. Kline and S. Hu, *Adv. Synth. Catal.*, 2003, **345**, 524.
11. C. A. Martinez, S. Hu, Y. Dumond, J. Tao, P. Kelleher and L. Tully, *Org. Process Res. Dev.*, 2008, **12**, 392.
12. D. R. J. Acke and C. V. Stevens, *Org. Process Res. Dev.*, 2006, **10**, 417.
13. R. C. Wheeler, O. Benali, M. Deal, E. Farrant, S. J. F. MacDonald and B. H. Warrington, *Org. Process Res. Dev.*, 2007, **11**, 701.
14. A. R. Bogdan and N. W. Sach, *Adv. Synth. Catal.*, 2009, **351**, 849.
15. (a) J. P. Tierney and P. Lidstrom, *Microwave Assisted Organic Synthesis*, Wiley-Blackwell, Oxford, 2005; (b) C. O. Kappe, D. Dallinger and S. Murphree, *Practical Microwave Synthesis for Organic Chemists: Strategies Instruments and Protocols*, Wiley VCH, Weinheim, 2008.

16. O. Benali, M. Deal, E. Farrant, D. Tapolczay and R. Wheeler, *Org. Process Res. Dev.*, 2008, **12**, 1007.
17. L. Kumar, A. Amin and A. K. Bansal, *Drug Discovery Today*, 2007, **12**, 1046.
18. J. F. Remenar, J. M. MacPhee, B. K. Larson, V. A. Tyagi, J. H. Ho, D. A. McIlroy, M. B. Hickey, P. B. Shaw and O. Almarsson, *Org. Process Res. Dev.*, 2003, **7**, 990.
19. N. G. Anderson, *Org. Process Res. Dev.*, 2004, **8**, 260.
20. L. C. Hsu, L.-J. Ping, E. C. Webb and T. J. Wrzosek, *Lab. Rob. Autom.*, 1999, **11**, 186.
21. S. M. Chesnut and J. J. Salisbury, *J. Sep. Sci.*, 2007, **30**, 1183.
22. C. Dinter, H. Weinmann, C. Merten, A. Schutz, T. Blume, M. Sander, M. Harre and H. Neh, *Org. Process Res. Dev.*, 2004, **8**, 482.

CHAPTER 4
Microwave Assisted Chemistry

RACHEL OSBORNE

Worldwide Medicinal Chemistry, Pfizer Ltd., Ramsgate Road, Sandwich, CT13 9NJ, UK

4.1 Introduction

Microwave assisted chemistry is a field which has grown in importance relatively recently in the synthetic organic community. Since the first reports from Gedye *et al.*[1] and Giguere, Majetich *et al.*[2] in 1986 which demonstrated the use of commercial domestic microwave ovens as the heat source for a range of common organic reactions, and with the rapid development in the mid-1990's of specialist equipment designed to increase safety and physical parameter monitoring, it can now be said that a synthetic laboratory is poorly equipped if it does not have microwave capabilities, such is the usefulness and efficiency of the technique.

  Modern synthetic organic chemistry routinely involves the heating of reactions and traditionally, this would involve the use of oil baths, isomantles and, more recently, oil-free options consisting of metal hotplate bases with round-bottomed flask-sized inserts. All of these options involve heating the reaction by transferring energy to the reaction vessel, then to the bulk solvent by conduction (from vessel wall to adjacent solvent) and convection (within the bulk solvent). By heating in this manner, much of the energy is used in purely heating the glassware and the reaction mixture itself is rarely found to be at the same temperature as the vessel wall, with the centre of the solution often significantly cooler than the outside.

RSC Drug Discovery Series No. 11
New Synthetic Technologies in Medicinal Chemistry
Edited by Elizabeth Farrant
© Royal Society of Chemistry 2012
Published by the Royal Society of Chemistry, www.rsc.org

Microwave assisted synthesis offers a different approach: by directing microwave irradiation into the centre of the vessel, more uniform heating can be achieved as the solvent is heated directly by the microwave energy, without the need to heat the glassware. The main advantage of microwave chemistry is that it generally allows reactions to be carried out much more quickly than thermal heating and the benefits of this are: increased efficiency, faster turn-around of reactions leading to increased productivity and higher purity of reaction mixtures.

Whilst an overview of the physical theory of microwave technology will be provided, it is not intended that this chapter will serve as an in-depth analysis of the field, as there are many excellent sources of such information.[3,4] Neither is it the intention for it to be a literature review of all reactions carried out under microwave conditions, as there are many books[3-6] and review articles[7-12] which the reader might consult for a more comprehensive coverage of the range of applicable chemistries.

Instead, this chapter aims to discuss the benefits of microwave assisted synthesis and a current analysis of the equipment available, along with some practical considerations and thoughts on the best strategy for its use. The examples shown will aim to give an overview of the range of chemistry which has been successfully conducted using microwave technology, with particular emphasis on reactions one might find commonly employed within a pharmaceutical laboratory and with a view to comparing the profiles of reactions carried out under microwave conditions *versus* those carried out thermally.

4.2   An Overview of Microwave Theory

4.2.1   How Do Microwaves Enhance Chemical Reactions?

In the early stages of research into microwave mediated reactions, there was considerable debate over how microwave energy enhanced chemical reactions. Although initially the idea was that microwave energy directly interacted with reactants thus promoting the reaction (non-thermal effects), this view has become less favourable over time (though it is still believed by some to be of significance for some reactions). Instead, there is a general belief that the enhancement in reaction outcome seen in microwave assisted reactions can be attributed solely to effects mediated by the heating of the bulk solvent (thermal effects), though there are examples (especially in heterogeneous reactions) whereby the direct heating of reagents contributes to the enhanced rate of reaction.

It should be stressed that there is still considerable debate over the nature of the "microwave effect" and that no single explanation can cover all reaction types. As it is outside the scope of this short review to give a thorough coverage of the subject, the interested reader is recommended to some of the excellent reviews and papers on the subject.[3,4,5,13-15]

Because of limitations of space, and as the general consensus has moved away from the thought that non-thermal effects are the norm, only thermal effects will be considered in this section.

4.2.2 Microwave Heating

Microwave heating of reactions is mediated by the interaction of the electrical component of electromagnetic radiation with the components of the reaction. This can be through either dipolar polarisation (for molecules with a dipole moment) or ionic conduction (for anions and cations).

Dipolar polarisation is the phenomenon whereby molecules with dipole moments attempt to align those dipole moments with the electric component of electromagnetic energy. As this component is continuously oscillating, realignment is continuous; in the process, heat is generated through intermolecular friction and collisions.

Ionic conduction is the phenomenon whereby ions oscillate within the electric field: collisions with other molecules or atoms in the sample generate heat. A good example of the difference this makes is in the microwave heating of a sample of tap water (containing a mixture of dissolved inorganic salts) and distilled water (which is essentially pure water): the tap water sample will heat faster than the distilled water sample (under the same power input).

The ability of a solvent to be heated by microwave energy is not solely dependent on its dipole moment, but upon its loss tangent, $\tan \delta$:

$$\tan \delta = \varepsilon'' / \varepsilon' \tag{4.1}$$

where ε'' is the dielectric loss (how efficiently microwaves are converted into heat) and ε' is the dielectric constant (how polarisable the molecules are in an electric field).

The distinction is important, as can be seen in Table 4.1, as some solvents with a large dipole moment (ε') (*e.g.*, water) are not as effectively heated as ones with a slightly lower dipole moment (*e.g.*, ethanol) but with a higher $\tan \delta$ value. It is still possible to use solvents with low $\tan \delta$ values either by adding a co-solvent with a higher $\tan \delta$ value or by adding salts or an ionic liquid.

The $\tan \delta$ values are not fixed but vary with both the frequency of electromagnetic radiation and the temperature. As most microwave reactors are of the same operating frequency, the former does not need to be considered, but it should be noted that $\tan \delta$ values generally decrease with increasing temperature.

The main advantages of microwave heating over conventional methods are:

1. The ability to easily achieve temperatures that are difficult to achieve (and maintain) using conventional methods;
2. Direct heating of solvent, not the reaction vessel; and
3. Lower energy requirements.

Table 4.1 tan δ values for a range of common solvents. Data from ref. 15.

Solvent	tan δ
Ethylene glycol	1.350
Ethanol	0.941
DMSO	0.825
2-propanol	0.799
Methanol	0.659
1-butanol	0.571
NMP	0.280
Acetic acid	0.174
DMF	0.161
Water	0.127
Chlorobenzene	0.101
Acetonitrile	0.062
Ethyl acetate	0.059
Acetone	0.054
Tetrahydrofuran	0.047
Dichloromethane	0.042
Toluene	0.040
Hexane	0.020

1. As a general rule, achieving temperatures in excess of 140–150 °C using conventional methods (oil bath or heating block) with standard laboratory glassware is difficult and often requires the insulation of the reaction vessel. The use of a sand bath and sealed vessel has allowed chemists to achieve higher temperatures but this method has very limited utility (especially as stirring is almost impossible). However, by using a microwave reactor it is very easy to achieve solvent temperatures of up to 200 °C for the vast majority of reactions and higher temperatures, up to the manufacturer recommended limit of around 250 °C, may be achieved in certain cases. Several solvents can be heated to temperatures in excess of their boiling points (sometimes by as much as 100 °C). Not only are such high temperatures achievable, they can be achieved rapidly in timescales that are almost impossible to reproduce by conventional heating (N-methylpyrrolidinone (NMP) can reach 200 °C in under a minute).

   One of the major advantages of the ability to run reactions at higher temperatures than would normally be feasible using standard laboratory conditions is the observed enhancement of the reaction rate. The rate of a chemical reaction can be described using the Arrhenius equation:

$$k = Ae^{-E_A/RT} \tag{4.2}$$

where k is the rate constant of the reaction, A is the pre-exponential factor, e is the standard mathematical quantity, E_A is the activation energy, R is the gas constant and T is the temperature in degrees Kelvin.

If A and E_A remain constant, then a ten degree increase in reaction temperature results in an approximately two-fold increase in reaction rate, or each time the reaction temperature is raised by ten degrees, the reaction time required is halved. For small changes in reaction temperature this has a moderate effect, but with microwave irradiation being able to easily heat reactions to temperatures in excess of 150 °C, significantly reduced reaction times can be achieved. For example, a reaction requiring 24 h to reach completion at 80 °C should only require about 1.5 min at 180 °C. It should be pointed out that such calculations are only approximate and ignore factors such as the time to achieve the reaction temperature and the efficiency of mixing, but they provide a good first approximation for the chemist converting a conventional method into a microwave assisted one.

2. Direct heating of the solvent, not the reaction vessel, results in a significantly lower reaction vessel temperature than in conventional heating. This reduces possible wall effects whereby unwanted side reactions occur at the solvent/reaction vessel interface. This is particularly important in heterogeneous reactions where, under conventional methods, catalyst decomposition on the hot reaction vessel wall is a major route of catalyst deactivation. Additionally, the solvent is heated more uniformly, reducing the possibility of localised hot spots within the bulk solvent.

3. Microwave heating is more efficient in terms of the total energy required to heat a reaction.[16] Studies comparing the energy efficiency of conventional heating methods and microwave assisted synthesis showed that, for the majority of reactions, significantly less energy is required when using microwave irradiation as the heat source. Although this difference is not large for small scale reactions, it becomes significant as scale increases and by heating the reaction using a microwave reactor one can improve the "greenness" of the reaction.

4.3 An Overview of Commercial Microwave Reactors

Microwave assisted chemistry began with the use of modified domestic microwaves in the 1980's, as described in the key papers by Gedye *et al.*[1] and Giguere, Majetich *et al.*[2] In these experiments, there was no accurate control of the power input, nor any monitoring of reaction parameters such as temperature or pressure. The microwave energy was not focussed on the centre of the vessel compared to modern apparatus and therefore the heating may not have been uniform. However, these experiments did show dramatic rate enhancements across a range of common transformations and led to an exponential growth in the number of subsequent publications in the area.

Microwave assisted synthesis has developed from those early days of using domestic microwaves, often modified with crude temperature probes or reflux condensers, which would be placed through a hole in the top of the domestic microwave. As we moved into the 21st century, commercial microwave technology reached a point where several companies produce a range of bespoke microwave reactors which are suitable for reactions from a small discovery scale up to process-scale reactors, with all the in-built safety devices and parameter monitoring functions that one would expect from a highly technical piece of laboratory equipment.

This section will give an overview of the general features of the equipment. For more specific information, the reader might like to consult one of the manufacturers' websites,[17-19] and also to refer to one of the excellent equipment reviews that have previously been published.[4,6]

Commercial microwave equipment for synthesis falls into two broad categories: single-mode (often termed mono-mode) and multi-mode. The single-mode apparatus delivers a "standing wave" of focussed microwave energy to a single vial which is positioned a fixed distance away from the magnetron and in this mode a single vial can be heated uniformly to a specific temperature. Many of the commercial single-mode machines are modular in design, in that it is possible to attach additional functionality to the basic model. These modules include automated robotic sample racks, whereby single vial experiments can be queued to be run sequentially, and external cooling devices (over and above the compressed air cooling system as standard) which enable the cooling of the sample during irradiation to maximise power input.

The multi-mode apparatus is designed to irradiate a larger volume than single-mode. In order to avoid localised hot-spots of focussed microwave energy (caused by standing waves), a mode stirrer deflects the incoming microwave energy making it reflect around the cavity in a chaotic and unfocussed manner. Multi-mode reactors are ideal for use in parallel chemistry; however, as there is still some uneven distribution of microwave energy within the cavity, they have a rotor arm and the samples are continuously rotated within the cavity to ensure a more uniform heating profile.

As can be seen in Figure 4.1, there is some variation between the equipment developed by each manufacturer; however, the fundamental design features of each machine are similar. Most of the discovery scale machines will fit into a standard size fume hood, making them easily integrated into the laboratory environment.

The upper temperature limit is usually around 250 °C, and most designs are capable of ramping the temperature up by between 0.5 and 5 °C s^{-1}, giving a very controlled heating mechanism. The cavity is designed to withstand pressures of up to 20–30 bar and all machines are equipped with safety locks and monitoring probes (IR sensors for temperature and integrated pressure sensors in the cavity lid) which regulate the power output and allow for a safety cut-off in case of a runaway reaction. Most reactors have in-built magnetic stirring capability, although some of the large scale reactors are equipped with mechanical stirring devices. They are all cooled by a pressurised air supply system, which allows for rapid cooling at the end of the reaction time. Figure 4.2 shows a generalised schematic of a single vial mono-mode microwave reactor cavity illustrating these design features.

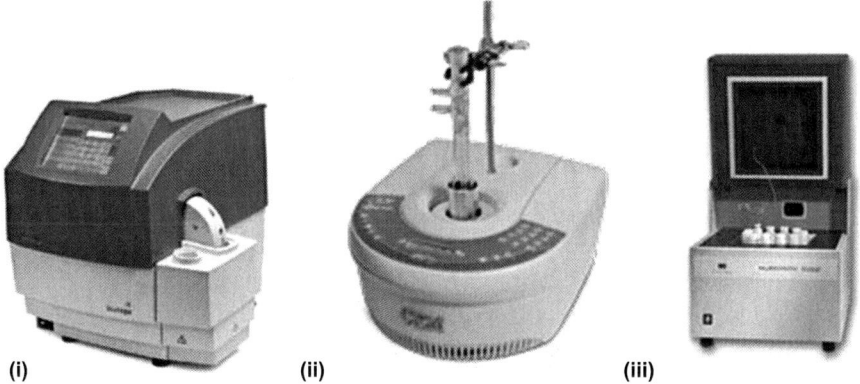

Figure 4.1 A selection of commercially available microwave reactors: (i) Initiator™ (single-mode), from Biotage AB; (ii) Discover® (single-mode) from CEM Corporation; (iii) MultiSYNTH (multi-mode) from Milestone Inc.

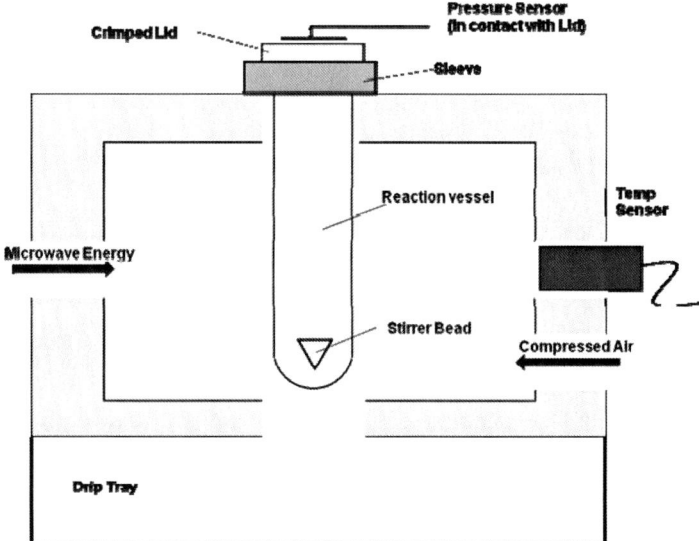

Figure 4.2 Generalised schematic of a single-vial, mono-mode microwave reactor cavity.

Most vials available from commercial suppliers are made from microwave-transparent materials such as borosilicate glass or quartz and are designed so as to allow the microwaves to pass through into the reaction mixture without energy loss to the vessel walls. The vials are designed to withstand pressures of up to 20 bar and are designed to accept a range of magnetic stirrers. They are equipped with caps that crimp on to the vial and bear a septum that serves to

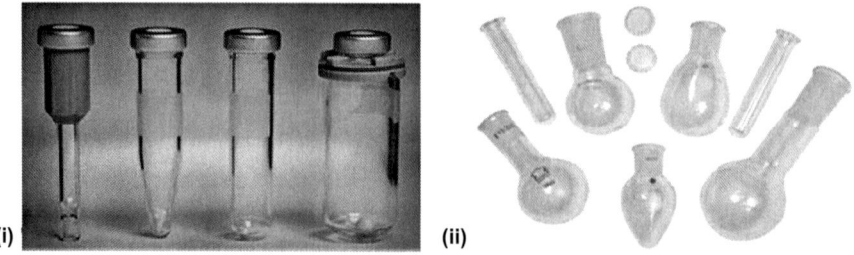

Figure 4.3   A selection of commercially available microwave vials: (i) Biotage AB, (ii) CEM Corporation.

allow firstly the injection of reagents, secondly the sampling of the reaction mixtures for analysis without the need to remove the cap, and thirdly the venting of the reaction vessel with a needle prior to decapping. As shown in Figure 4.3, manufacturers' ranges comprise vials and glass vessels in many different sizes to allow reactions on scales from milligrams to grams.

Some microwave reactors are convertible to open vessel technology, and have a port in the roof of the machine through which a reflux condenser can be attached to the vessel. Manufacturers of microwave reactors supply specialist glassware with standard quick-fit adaptors for use in open vessel reactions. In cases where this setup is used, the achievable temperature within the reaction is usually lower than that achievable in the sealed vial.

Many manufacturers also have ranges which include scale up or process synthesisers thus allowing the transfer of a reaction from small scale to large scale without the requirement for extensive reaction optimisation. These fall into two main categories: batch reactors and flow reactors. Batch reactors are capable of reactions on a litre scale, with large vessels and large microwave cavities. Flow reactors offer the benefit of a smaller reaction volume at any given point in time, and therefore have a smaller microwave cavity. They take advantage of recent developments in flow technology to allow for a continuous flow of material through the irradiation process, which can result in large scale production in short periods of time.

4.4   Current Applications of Microwave Assisted Synthesis

As mentioned earlier, it would be impossible for this chapter to be a thorough literature review of microwave chemistry given the huge size of the field. Instead, examples will be shown where microwave assisted chemistry has enabled the synthesis of specific compounds, or groups of compounds, with a view to illustrating and reinforcing the previously mentioned benefits of microwave usage. The examples will range from singleton chemistry to parallel-enabled library chemistry to process-scale chemistry, and aim to show the breadth of the applicability of the technique.

From personal experience it can often be the case that microwave chemistry is seen as a "last resort", to be used when all attempts at a classical thermal reaction have failed. The reality is that microwave-assisted chemistry should be the "first resort". It should be used to conduct all the trial reactions necessary to determine whether a reaction will work, be it a reaction screening exercise or a catalyst or solvent selection decision. By running the trial reactions in the microwave, the chemist benefits from increased efficiency and more timely decision making, allowing the scale up to commence on the same day without the need to wait overnight to determine if the trial reaction has worked or not.

If a reaction has worked *via* thermal heating, then it will work under microwave conditions. Microwave chemistry is, however, not a magic "cure-all", in that it generally will not make a reaction work when classical thermal heating has failed. To qualify that point, if prolonged heating at high temperatures has failed thermally, there would usually be a low possibility of microwave heating changing the situation. However, it is important to understand whether there has been any change in the reactant profile. If the reagents are unchanged, then the previous point holds true. However, if there is evidence that the lack of desired product is due to degradation of that desired product due to the long reaction time, then microwave technology may be able to improve the profile by reducing the length of the reaction time.

4.4.1 Metal-catalysed Reactions

Metal-catalysed reactions have been at the forefront of microwave assisted synthesis since the development of the technique and there are many examples of microwave assisted cross-coupling reactions in the literature.[20,21] Biaryl motifs are common in drug molecules[22] as they are known to interact favourably with various polar moieties such as amides and hydroxyl groups. The formation of these motifs is therefore of importance to the synthetic chemist. Most methods of synthesis rely on the coupling of an aryl halide or pseudohalide (for example, a triflate) and an organometallic reagent, for example, boronate (Suzuki–Miyaura), zincate (Negishi), stannane (Stille) or Grignard reagents (Kumada).

The ubiquitous Suzuki–Miyaura cross-coupling is well investigated under microwave conditions, and a representative example can be found in a paper by Cao *et al.*[23] The cross-coupling of a chloropyridazinone with a range of boronic acids is investigated under thermal and microwave heating conditions as can be seen in Scheme 4.1. From initial investigations, it had been observed that dehalogenation of the starting material was a major concern, leading to poor yields and difficult purifications. To facilitate the process of identifying a set of conditions, a reaction screening exercise was undertaken in which a number of different catalysts, ligands, bases, and solvents were trialled, each being irradiated for 30 min, to test the concept rapidly. This process established a best set of conditions which were then used to synthesise a range of products in moderate to excellent yields and, crucially, without the complication of the dehalogenation by-product.

More recent papers have discussed the use of microwave heating to make the coupling partners, for example boronic esters or zincates, and then their use in subsequent reactions.

The formation of a boronic ester from an aryl halide and a diboron species is standard methodology, although many of the boronates that are formed are not very stable, especially those on heteroaromatic rings. Also, these reactions are notorious for their complex product profile leading to lengthy purifications and a paper by Rheault et al.[24] demonstrates this point well. When synthesis of an aryl boronate did not proceed well thermally (14 h at 80 °C, giving 20% conversion to the boronate), microwave conditions were investigated as depicted in Scheme 4.2. It was discovered that heating the reaction mixture in the microwave for 20 min at 150 °C gave 45% conversion to the boronate species. The conversion increased to 59% after 20 min at 170 °C, suggesting that flash heating is the key to this reaction. Interestingly, and to illustrate the point about degradation of the product, it was shown that microwave heating at 150 °C for 40 min caused a slight 5% decrease in the yield of the reaction.

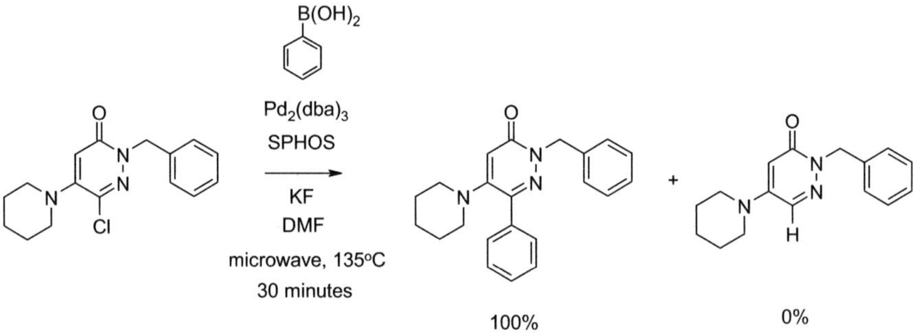

Scheme 4.1   Suzuki–Miyaura cross-coupling of a chloropyridazinone with boronic acid under microwave heating conditions.

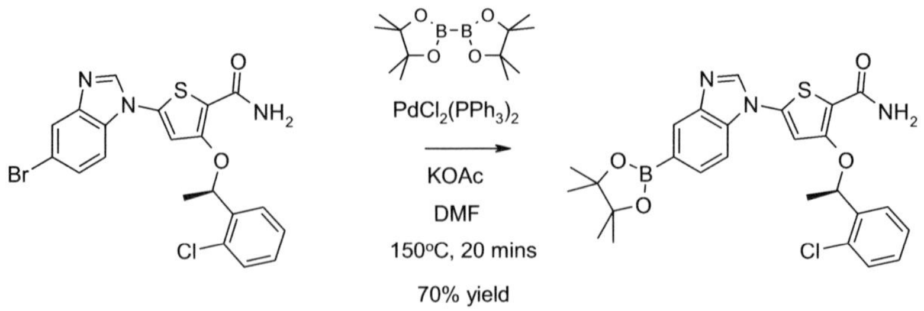

Scheme 4.2   Boronic ester formation under microwave conditions.

This suggests that prolonging the time that the boronate is heated has a deleterious effect on its formation.

Aryl zincates and aryl magnesiums have been synthesised under microwave conditions as described in a paper by Walla and Kappe.[25] The organometallics have then been used, again under microwave conditions, in cross-coupling reactions as shown in Scheme 4.3. The preparation of the zincates is quoted as taking several hours under reflux conditions by the conventional heating methodology, but needing only 5–30 minutes at 180 °C in the microwave using Rieke zinc dust. All reactions showed a conversion of >99% by HPLC or GC-MS. The preparation of organomagnesium reagents is also described in this paper, and the subsequent use in Kumada coupling is very high yielding as seen in Scheme 4.4.

A further example of organomagnesium generation is described in a paper by Nilsson *et al.*[26] In this instance, the product is used in a typical Grignard addition to benzaldehyde. Aryl Grignards can sometimes be difficult to form, and the electron-rich aryl chlorides used in this paper were shown to be sluggish under conventional heating, with low conversions and yields. Under microwave conditions, the yields of the Grignard products range from 89–99%, suggesting that the organomagnesium species has been formed in appreciable amounts as shown in Scheme 4.5.

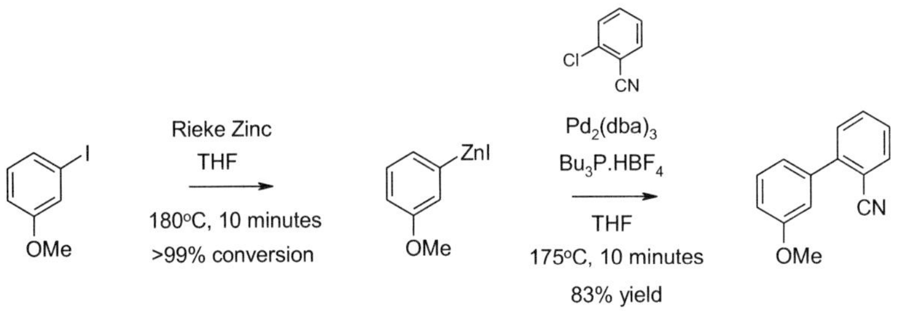

Scheme 4.3 Formation of organozincate and subsequent Negishi coupling under microwave conditions.

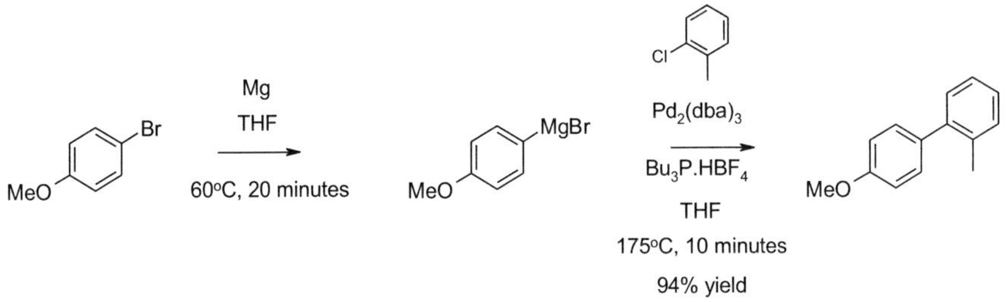

Scheme 4.4 Formation of organomagnesium species and subsequent Kumada coupling under microwave conditions.

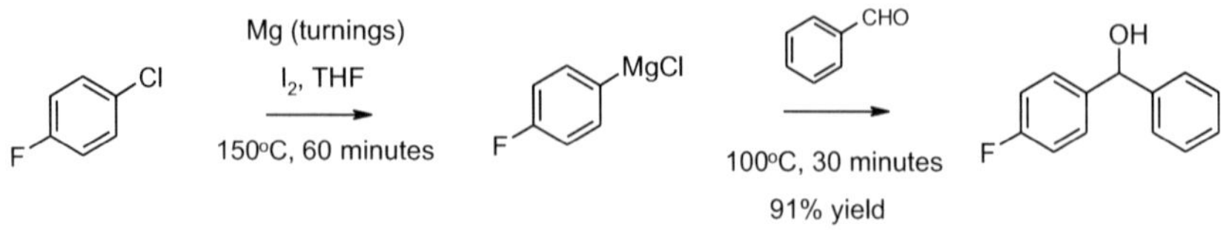

Scheme 4.5   Formation of organomagnesium species and subsequent Grignard reaction with benzaldehyde.

The above examples lead to an interesting practical consideration concerning the use of insoluble materials in the microwave. It is possible to run reactions where some of the material is not in solution in the case of inorganic bases or metal catalyst complexes. However, care should be taken to run the reaction at a reasonable dilution and to ensure that the particulates are washed off the sides of the vessel so as to be under the level of the solvent. This should prevent the particulates from becoming superheated and potentially creating "hot spots" on the vessel wall, which can lead to explosion of the vessel. In cases where there is an amount of insoluble material in the vessel, a pre-stirring time of around 10–20 s is usually recommended by most manufacturers.

4.4.2   S$_N$Ar reactions: Rate and Purity Enhancements

The increased rate of reaction observed when using microwave heating can best be demonstrated by looking at examples where conventional and microwave reactions have been compared. In the paper from Gedye *et al.*[1] in 1986, several organic transformations were studied both thermally and under microwave conditions. In all cases a rate increase was observed, from a 5-fold improvement for an amide hydrolysis to an excellent 240-fold increase for an S$_N$2 alkylation reaction. The main observation from the work was the inverse relationship between rate enhancement and the boiling point of the solvent used. That is to say, the lower the boiling point of the solvent, the more dramatic the rate increase. Equally, the higher the boiling point of the solvent, the smaller the rate increase.

The main benefit of decreased reaction times to the industrial chemist is the ability to perform more reactions in a fixed time period thereby being more productive. If those reactions also benefit from the often observed improved purity profile seen with microwave assisted reactions, then even more time is saved by avoiding lengthy work-up and purification procedures.

A good example of this can be seen in a paper by Harbottle *et al.*[27] where the microwave assisted synthesis of a series of carbon-, nitrogen- and oxygen-linked derivatives of pyrido[3,4-d]pyrimidin-4-ylamines is described. Amino-heterocycles are prevalent in drug molecules, and the most common method for their synthesis is *via* an S$_N$Ar displacement of a halogenated heteroaromatic. In the case of the above mentioned paper, the fluoride displacement reaction at the heart of the synthetic route was proving problematic under conventional

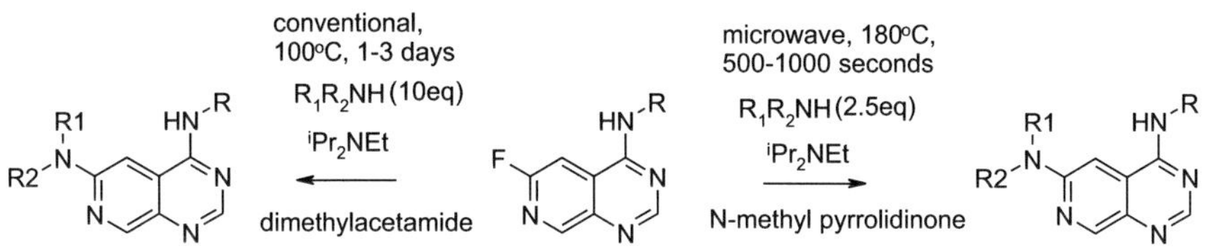

Scheme 4.6 A comparison between conventional and microwave heating in S_NAr reaction with amines.

heating conditions, with reactions taking between 1 and 3 days at 100 °C, using high excesses of the nucleophilic amine and producing product mixtures that required lengthy chromatographic purification. As can be seen in Scheme 4.6, by converting to microwave heating it was possible to reduce the length of the reaction time to between 10 and 30 min (at 180 °C) and to reduce the excess of the amine to 2.5 equivalents (this proved an additional advantage when using non-commercial amines). Not only was the reaction time dramatically reduced, but the crude reaction mixture was of sufficient purity to allow rapid isolation of the required product, whereas under conventional heating the products had to be purified by lengthy chromatography. The combined effect of both reduced reaction time and reduced purification time increased the productivity of the synthetic team, enabling them to create analogues of their lead compounds much faster than before.

4.4.3 Reactions Utilising a Gaseous Reagent

Microwave assisted synthesis with gas reagents is a recent development in the field. It has been postulated that the microwave vial can be thought of as a mini autoclave and as such could be used, with minor modifications, with gaseous reagents.[28] The classes of reaction one might consider include carbonylation (insertion of CO) and hydroformylation (formylation of terminal alkenes).

Microwave assisted carbonylation reactions have, in the past, mainly employed the molybdenum complex, $Mo(CO)_6$, which is a solid reagent and decomposes to release carbon monoxide under reaction conditions. However, it is highly toxic and on scale-up would result in large amounts of heavy metal waste, an important environmental consideration. The use of gaseous carbon monoxide would overcome these problems, and has been reported by Petricci *et al.* in a 2010 paper.[29] With the advances in commercial microwave technology it is now possible to introduce gases into the reaction tube, and carbonylation chemistry has been conducted under mild conditions as depicted in Scheme 4.7. Using only 9 bar pressure of CO gas, it has been shown that aryl iodides can be carbonylated in the presence of alcohols to give esters in very high yields. The paper also demonstrates that, under similar conditions, carbonylation in the presence of amines will generate amides in moderate to high yield. As can be seen in Scheme 4.8, a further useful application of this process is the cyclocarbonylation of iodo-anilines to give benzoxazinones, a useful scaffold in the pharmaceutical industry.

The hydroformylation of alkenes is a versatile method for the preparation of aldehydes. This metal-catalysed addition of carbon monoxide and hydrogen to a terminal alkene is a process used extensively in the field of speciality chemicals, fragrances and detergents and it is also a versatile addition to the synthetic organic toolkit within the pharmaceutical industry. Microwave conditions for this process have been developed as seen in Scheme 4.9 and reported in a paper by Taddei *et al.*[30] A modified vial equipped with a gas inlet was charged with the reagents and then subjected to microwave heating for 4 min. The internal pressure in the tube was monitored and a decrease was observed, indicating

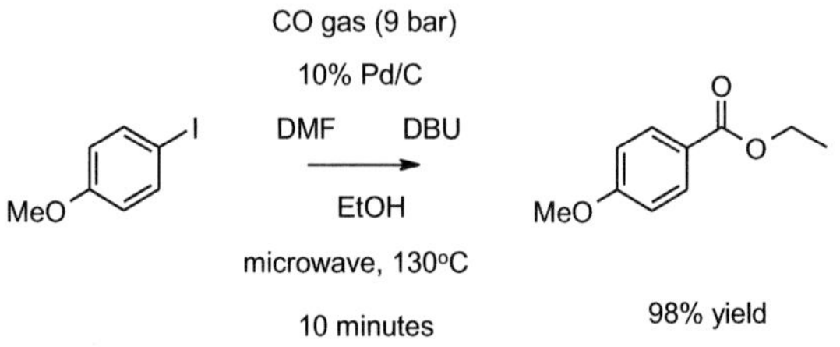

Scheme 4.7   Carbonylation under microwave conditions using CO gas.

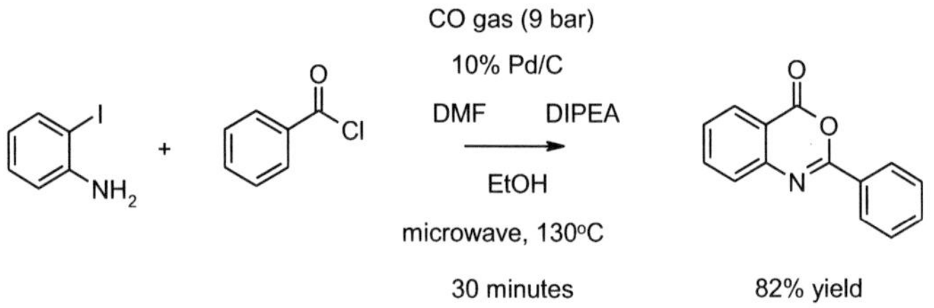

Scheme 4.8   Cyclocarbonylation under microwave conditions using CO gas to synthesise benzoxazinones.

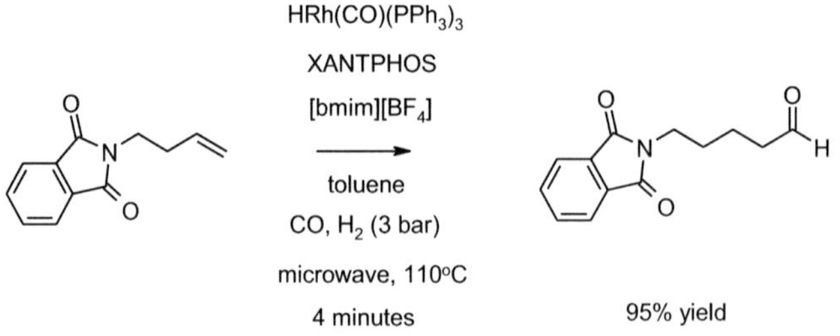

Scheme 4.9   Hydroformylation of terminal alkenes under microwave conditions.

that the reaction was proceeding. A range of terminal alkenes were subjected to the methodology and gave 70–90% yields.

4.4.4 Transfer Hydrogenations

One transformation where there has not been much published work is hydrogenation under microwave conditions using gaseous hydrogen. Alternatives to the use of gaseous hydrogenation have traditionally included transfer hydrogenation, using a hydrogen donor such as isopropyl alcohol, ammonium formate or 1,4-cyclohexadiene. These reactions are usually reliable but can be problematic with groups which resist hydrogenation and have been known to damage other sensitive functionality within the molecule due to the longer reaction times required. Transfer hydrogenation reactions have been studied under microwave conditions and have been shown to proceed quickly and without the associated by-products that can occur in thermally heated reactions; one example where this has been described is in a paper by Bäckvall *et al.*[31] A range of diketones is reduced *via* transfer hydrogenation with a ruthenium catalyst in isopropanol to form the corresponding diols in good to excellent yield as shown in Scheme 4.10. Previous attempts to reduce these 1,3 diketones with sodium borohydride had proved problematic due to the preference of the starting material to adopt the enol form. Microwave assisted hydrogenation using isopropanol as the hydrogen donor resulted in good isolated yields of the diol, with only acetone as a by-product. This paper is also a good example of running microwave reactions in low tan δ solvents such as toluene. It was possible to achieve the desired reaction parameters in this case due to the excess of isopropanol in the system, both as reactant and co-solvent.

The use of gaseous reagents in sealed vessels brings with it some safety concerns. A vial under pressure is a potential explosion risk and the experiment should be carefully planned to minimise such hazards. The level of the solvent should be minimised within the vial, so as to allow a large enough "head space" for the gaseous reagent. When planning a reaction, thought should also be

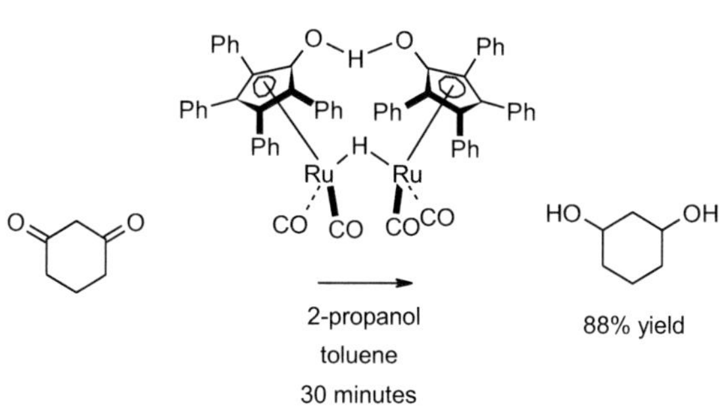

Scheme 4.10 Transfer hydrogenation of cyclic diketone with ruthenium catalysis and hydrogen donor.

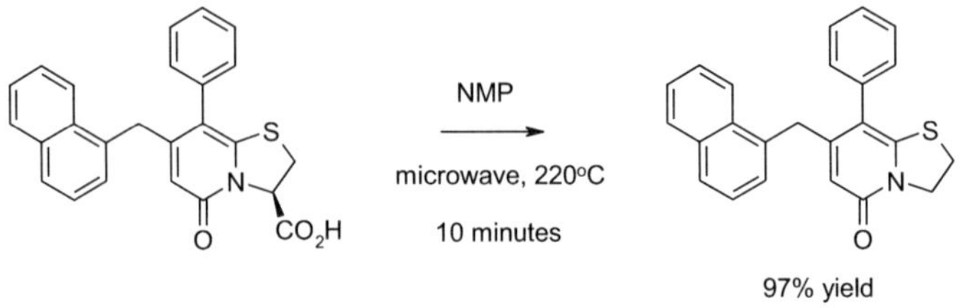

Scheme 4.11   Decarboxylation of bicyclic 2-pyridone under microwave conditions.

given to whether a gas could be formed in the course of the reaction, usually as a by-product, as this could bring potential explosion risks if there is not sufficient head space in the vial to accommodate the gas.

Thermal stability of some protecting groups, for example the *tert*-butyl carbamate (BOC) group, needs to be taken into account when planning a synthetic route. The BOC group can be thermally removed at temperatures around 180 °C[32] but begins to decompose at 140 °C,[33,34] to release carbon dioxide and isobutylene (both gases), thereby raising the pressure inside the sealed vessel and with it bringing the potential for explosion. Compounds with potential for decarboxylation should also be carefully handled in the microwave, as they may also release gas during the heating process. A paper by Almqvist *et al.*,[35] where microwave heating at 220 °C for 10 min in NMP provided an efficient method for the decarboxylation of a bicyclic 2-pyridone carboxylic acid as illustrated in Scheme 4.11. Examples such as this show that planned generation of gaseous by-products can be undertaken in a safe and high yielding manner.

4.4.5   Synthesis of Aromatic Heterocycles

The synthesis of heterocycles in medicinal chemistry is of key importance, as the vast majority of drug candidates contain at least one such ring system. These can include multi-substituted single rings, or bicyclic fused systems and these often form the scaffold at the centre of a molecule onto which the other groups are attached. Methods of synthesis are numerous; however, they often require long reaction times and yields are variable. Many traditional heterocycle formations have been successfully transferred to a microwave assisted approach and in the process have seen increased yields and decreased reaction times.[36,37]

One such example can be seen in a paper by Guo *et al.*[38] Whilst working on a series of analogues of MCH-1 receptor antagonists, it was necessary to build a bi-heterocyclic core as the key step in the synthesis. The conventional heating method (AcOH, 120 °C, 15 h) gave no detectable amount of the desired product; however, on transfer to the microwave protocol, (AcOH, 200–220 °C, 200–600 s) yields of 44–89% were achieved, an example of which can be seen in Scheme 4.12.

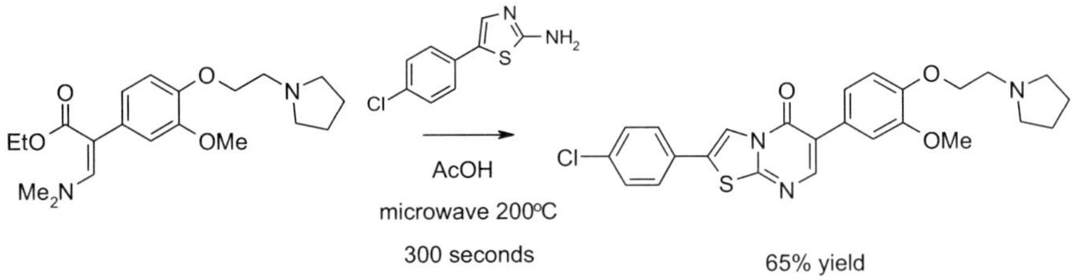

Scheme 4.12 Heterocyclisation of dimethylaminopropenoates to give bicyclic core.

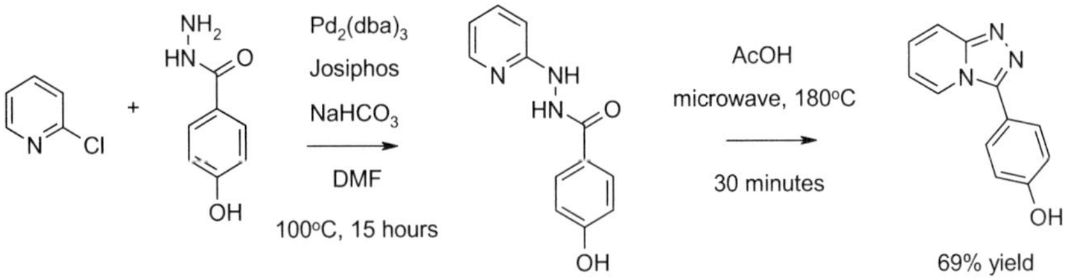

Scheme 4.13 Dehydration of acyl hydrazide intermediate to yield triazolopyridine core.

The synthesis of fused bicyclic scaffolds is also the concern of a paper by Reichelt *et al.*[39] in which the [1,2,4]triazolo[4,3-a]pyridine core is made in moderate to excellent yield. This core is found in many bioactive molecules, including antibacterial and anti-inflammatory compounds. The formation of the intermediate pyridyl acyl hydrazide from 2-chloropyridine and a range of benzoic hydrazides was accomplished thermally using palladium catalysis in excellent yield. The intermediate was isolated and then subjected to cyclisation conditions. This was investigated thermally and by microwave irradiation. The thermal method involves heating in polyphosphoric acid (PPA) at 110 °C for 15 h, resulting in complete conversion and 82% isolated yield. However, the reaction time is quite long, and the work-up of PPA-mediated dehydrations can be difficult due to its viscosity and water reactivity. When microwave conditions were investigated as illustrated in Scheme 4.13, it was discovered that the cyclisation would proceed in quantitative yield in acetic acid at 180 °C for 30 min. Not only has the reaction time been dramatically reduced, the work-up is also much simplified.

One further example can be seen in a paper by Besson *et al.*,[40] which studies the microwave assisted synthesis of Azixa™, a compound currently in clinical trials for the treatment of brain tumours. The compound can be made in two steps, both using microwave heating *via* a Dimroth rearrangement to yield a quinazoline core. As can be seen in Scheme 4.14, the initial reaction of 2-aminobenzonitrile with dimethylacetamide dimethylacetal proceeds in as little as 2 min under microwave conditions in 90% yield. Under Lewis acidic

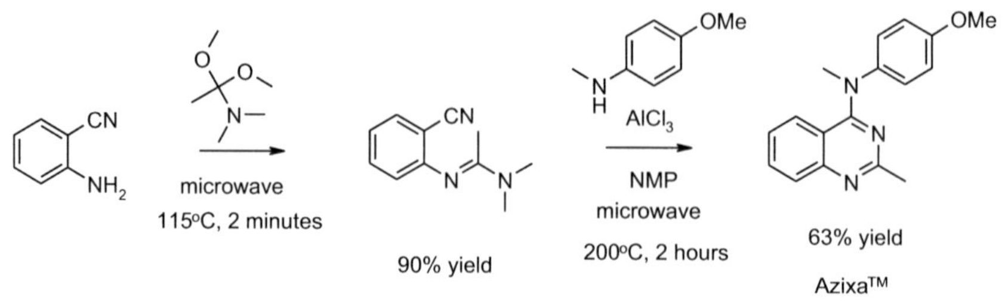

Scheme 4.14 Dimroth rearrangement in the synthesis of Azixa™ under microwave conditions.

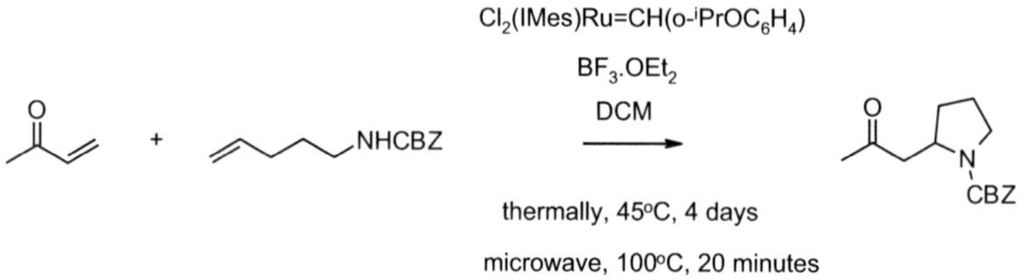

Scheme 4.15 Thermal *vs.* microwave in the cross-metathesis/aza-Michael addition reaction to form a cyclic β-amino acid functionality.

conditions, the Dimroth rearrangement is achieved in 63% yield to yield Azixa™ by an alternative to the existing manufacturing route.

4.4.6 Synthesis of Saturated Heterocycles

Microwave technology has also been used in the synthesis of saturated heterocycles such as pyrrolidines and in the ring formation of lactams.

One class of compounds where pyrrolidines are found is the cyclic β-amino acids, a useful building block both in small molecule drug targets and in the synthesis of modified peptides. In a paper by Fustero *et al.*,[41] cyclic β-amino carbonyl derivatives are synthesised *via* a novel cross metathesis/aza-Michael addition of vinyl ketones and amino alkenes. Use of the Hoveyda–Grubbs metathesis catalyst facilitates the tandem cross metathesis reaction followed by the aza-Michael addition to the amine to form the cyclic β-amino ketone as can be seen in Scheme 4.15.

Comparison of the reaction thermally and under microwave conditions shows that the thermal reaction takes 4 days at 45 °C, whereas the microwave mediated reaction takes only 20 min at 100 °C. The yield is the same in both cases, showing that this is just a rate enhancement through microwave irradiation. However, an interesting effect is seen when the amine starting material is α-substituted with various groups. When heating the reaction thermally, after 4 days, the *trans* (relative) diastereomer predominates. However, when heating under microwave conditions, the reverse was observed with the major dia-stereomer having the *cis* disposition as illustrated in Scheme 4.16.

Lactams are well precedented in drug targets, most notably the β-lactam antibiotics, but the larger ring sizes are also of synthetic interest to medicinal chemists. Lactams including pyrrolidinones and piperidinones have been syn-thesised by an Ugi 3-component reaction under microwave heating conditions in the absence of solvent as depicted in Scheme 4.17. In a paper by Deprez *et al.*[42] the Ugi condensation of a keto-acid, an isocyanide and an amine was studied, varying all three groups and the microwave conditions to give 5- and 6-membered lactam rings in excellent yields. The 5-membered pyrrolidinone ring has been studied in this way in previous papers, but this was the first example of a 6-membered piperidinone ring formation under microwave conditions. The yields ranged from 80–94% across a range of substituted products, giving an increase in yield compared to the conventionally heated results.

4.4.7 "Click Chemistry"

A hot topic in recent years within the synthetic community is the field of "click chemistry": the reaction of two chemical components in high yield with high atom efficiency in a simple and clean process. There are many excellent reviews

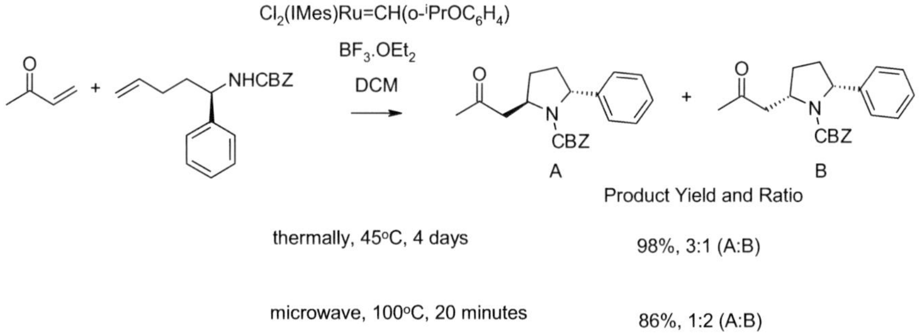

		Product Yield and Ratio	
thermally, 45°C, 4 days		98%, 3:1 (A:B)	
microwave, 100°C, 20 minutes		86%, 1:2 (A:B)	

Scheme 4.16 Selectivity inversion in the comparison of thermal *vs.* microwave in the cross-metathesis/aza-Michael addition reaction of α-substituted amines to form a cyclic β-amino acid functionality.

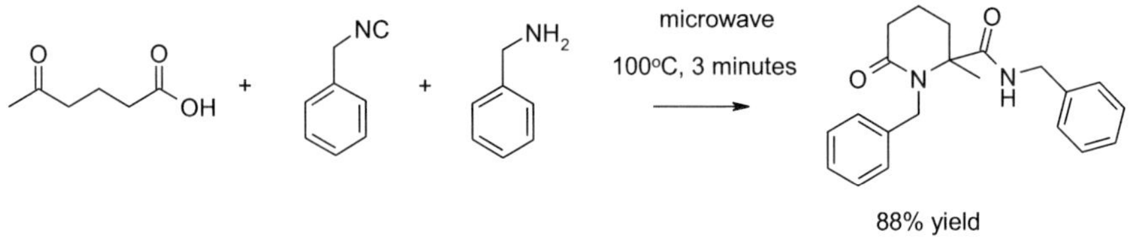

Scheme 4.17 Ugi 3-component condensation to yield lactam under microwave conditions.

of the area which the reader might consult for a more in-depth view of the ranges of chemistry that are encompassed by the click philosophy.[43–45]

The click reaction that has been most extensively studied is the Huisgen [3 + 2] dipolar cycloaddition. This is the reaction in which an azide (either pre-existing, or made *in situ*) reacts with an alkyne triple bond in a copper-mediated process to form a 1,2,3-triazole. It is now often referred to as a copper-catalysed azide–alkyne cycloaddition (CuAAC). It is often the case that the triazole is formed as a linking group between two halves of a molecule, and it is often one of the later steps in a synthetic route. Triazoles are often seen in bioconjugation chemistry, acting as a link between the biological fragment (for example, protein or DNA) and the chemical fragment (for example, fluorescent marker or small bioactive molecule). Whilst the cycloaddition works very well under thermal conditions, the benefit of using microwave energy to effect the transformation is the reduced reaction time, which means less prolonged heating of the very high energy azide and alkyne functional groups and also the potentially unstable bioconjugate.

A representative example of this cycloaddition is found in a paper by Moorhouse and Moses[46] concerning a one-pot azide formation and triazole formation. A range of anilines are converted *in situ* to azides using t-butyl nitrite and TMS-azide. The resulting intermediate azide is then reacted with a range of alkynes to yield the triazole, an example of which is shown in Scheme 4.18. Under conventional conditions at room temperature overnight, a 65% yield was obtained. In the microwave reactor, the reaction was heated at 80 °C for 10 min and gave a 97% yield. The temperature-sensitive functionality present in the starting materials meant that it was not possible to raise the temperature further and longer reaction times did not seem to have a positive effect on lower yielding examples.

The tetrazole group has often been used within medicinal chemistry targets as a carboxylic acid bioisostere as they are comparably acidic and metabolically more stable. The ring formation of a tetrazole is accomplished *via* a [3 + 2]-cycloaddition between an azide and a nitrile. An interesting study of one-pot tandem reactions in a paper by Shie and Fang[47] shows the conversion of an aldehyde to a nitrile and then on to the tetrazole in aqueous media. The aldehyde is oxidised with iodine in ammonia/water at room temperature with excellent conversion. As can be seen in Scheme 4.19, the reaction mixture is

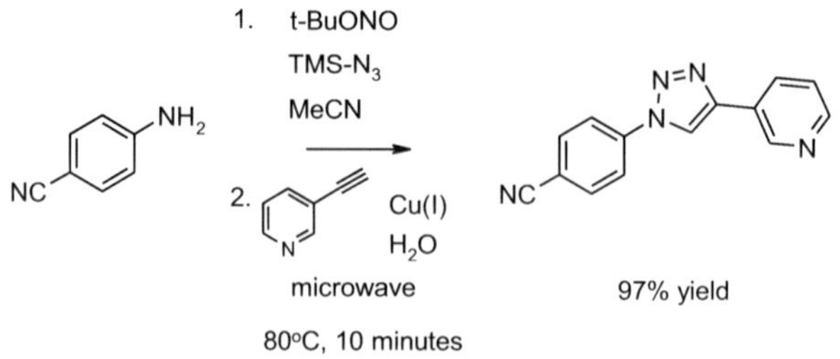

Scheme 4.18   Huisgen cycloaddition of azide and alkyne to give a 1,2,3-triazole.

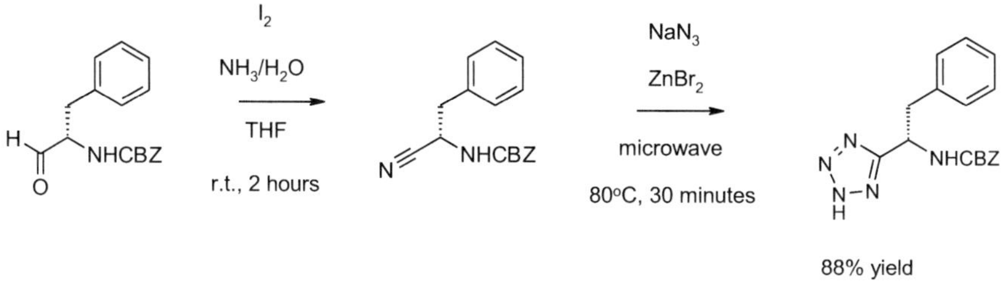

Scheme 4.19 *In situ* oxidation of aldehyde to nitrile, then cycloaddition with azide to furnish tetrazole.

then directly treated with sodium azide and zinc bromide prior to heating under microwave conditions for 30 min at 80 °C. The yields are excellent across the range of examples and, compared to the thermal heating method of prolonged reflux (17–48 h), microwave heating has offered a safer, greener and more efficient approach.

4.4.8 Reactions Utilising Solid-supported Reagents

Whilst the use of microwave heating has often been observed to increase the purity of the reaction mixture, it is possible to take that one step further with the use of polymer-supported reagents. These are everyday reagents that have been supported on resin beads and where they are used the reaction occurs at the solid/solution interface. The benefit is that the polymer, and with it the reagent, is simply filtered away from the reaction mixture at the end of the reaction. These reagents have been investigated in association with microwave heating and have been shown to be as stable and amenable as with thermal heating. As with any solids used in the microwave, care should be taken to ensure that the beads are below the solvent line and stirring freely. A paper by Wang *et al.*[48] shows the use of polymer-supported triphenylphosphine (PS-PPh$_3$) to effect the cycloaddition of amidoximes and carboxylic acids to give 1,2,4-oxadiazoles as illustrated by Scheme 4.20. As a useful bioisostere for amides or esters, 1,2,4-oxadiazoles have been found in many drug targets, but can be troublesome to synthesise. This study shows that by way of *in situ* conversion of the acid to the acid chloride using PS-PPh$_3$ and tri-chloroacetonitrile, followed by base-mediated cyclisation to the oxadiazole, it is possible to overcome the handling issues of using preformed acid chlorides and improve the yield and purity of the reaction.

4.4.9 Parallel Synthesis

As discussed in Chapter 2, making libraries of compounds is an efficient method for the generation of structure–activity relationships (SAR) within a medicinal chemistry program. By making the compounds in parallel, the design

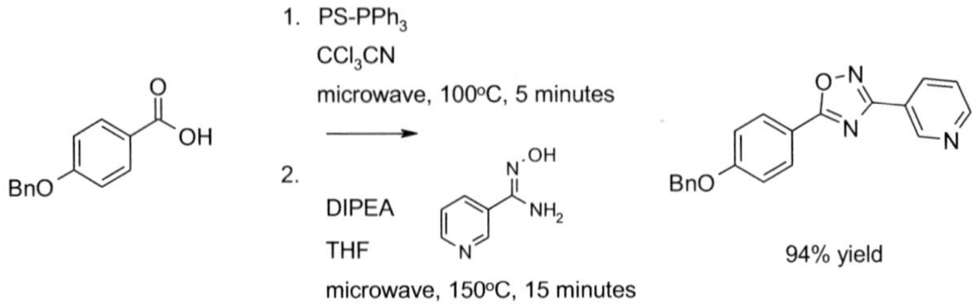

Scheme 4.20 Use of polymer-supported PPh₃ in the microwave assisted synthesis of 1,2,4-oxadiazoles.

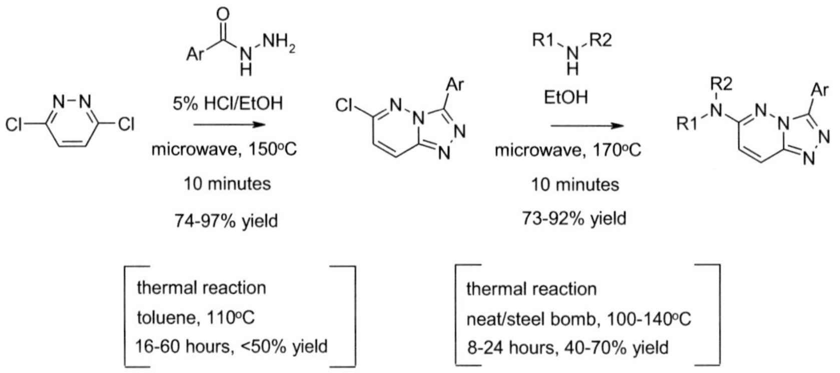

Scheme 4.21 Two-step protocol using microwave heating for both steps in an M_1 antagonist targeted library.

hypothesis can be tested more quickly. This process can be further accelerated by use of microwave assisted synthesis. Parallel reactions can either be run in a single vial in an automated sequential fashion or, with recent advances in the commercial technology, multiple reactions can now be run at the same time within a multi-mode apparatus.[49,50]

One example where the use of the microwave has had a direct impact on the progression of a project is described in a series of papers by Lindsley *et al.*[51,52] The synthesis of an M1 antagonist targeted library of 3,6-disubstituted-[1,2,4]triazolo[4,3-b]pyridazines over two steps (heterocyclisation followed by S$_N$Ar with amines) was difficult under conventional conditions. The first step of cyclisation with an acyl hydrazide was low yielding and took up to 60 h. The second step involved heating the amine and the chlorinated heterocycle under solvent-free conditions for up to 24 h in a steel bomb under pressure. The yields for this step were low. When the chemistry was adapted for microwave heating it was discovered that it was possible to acid-catalyse the initial ring formation and reduce the reaction time to 10 min as can be seen in Scheme 4.21. With the chloroheterocycle in hand, displacement with amines was investigated. Solvent was added, in this case ethanol, and the reaction was heated at 170 °C for

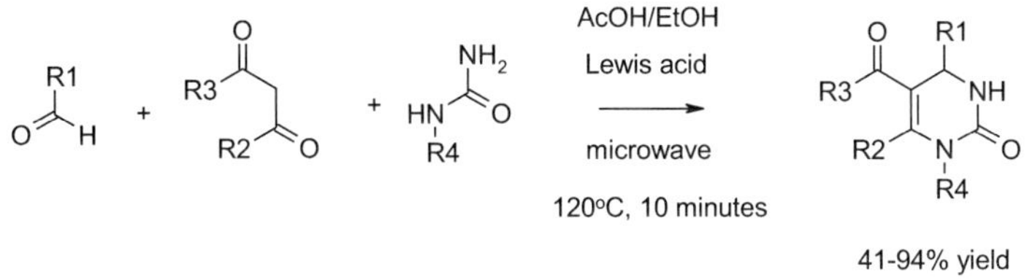

Scheme 4.22 Biginelli three-component condensation carried out in automated library format under microwave heating conditions.

10 min. Overall, multiple analogues of the lead compound were generated in only 20 min (plus isolation) compared with several days for a single analogue conventionally. In addition, the reaction is safer, as it no longer requires solvent-free, pressurised conditions.

In order to achieve the maximum benefit of using microwave chemistry to reduce reaction time, it is possible to automate the whole process of library synthesis. In a paper by Stadler and Kappe,[53] a fully automated process is described for the production of a 48 compound array *via* the Biginelli three-component condensation as shown in Scheme 4.22. Each of the reactants was made up as a stock solution and, using liquid-handling robots, were dispensed into the microwave vials *via* the septa. Each vial was then subjected to microwave irradiation for 10–20 min at 120 °C in an automated series fashion. Taking into account the time taken for processing functions (dispensing reagents, robot movements) it was possible to generate a 48 compound library in 12 h and, since in this case the products crystallised straight from the reaction mixture, the cycle time was very rapid.

4.4.10 Multi-gram Scale Reactions

Microwave assisted chemistry on small scale is well documented in the literature and, as described in all of the previous examples, has allowed for greater productivity within the discovery laboratory. However, for the technique to be of wider applicability and greatest benefit, there needs to be a method for scaling up reactions to kilo-lab scale and beyond.[54] There are many factors to take into account when attempting to scale up a reaction under microwave conditions. A short review by Kappe *et al.*[55] discusses these points, focussing on the various limitations of microwave use at large scales and giving case studies of the scale-up of several classes of organic reactions. The main concern with a large scale reaction is lack of microwave penetration to the core of the sample, as microwave energy may only reach the outer few centimetres of the sample within the vessel and result in uneven heating. This greatly limits the size of the vessel that can be used and, with it, the maximum batch size of the reaction. Large scale synthesis in batch mode requires the sequential heating of batches of reagents, followed by a purging of the apparatus before the next batch commences. The recent developments in flow chemistry technologies as described in Chapter 5 have been coupled with microwave technology to create a system which can allow for large scale synthesis in flow mode, whereby the vessel is fairly small but the continuous nature of the flow process means that large amounts of material can be generated in relatively short timeframes. There are many good examples of when flow technology has been used to enable the scale-up of a microwave reaction in a continuous process.[56,57] (Also, consult relevant references in Chapter 5 for further examples.) However, not all reactions will be suitable for scale-up in this fashion. Continuous flow reactors can become blocked with the solid material found in heterogeneous reactions, or with high viscosity liquids. In cases such as these, it may be more practical to use a batch process.

In a paper by Pitts *et al.*,[58] the scale-up of the final step in the synthesis of citalopram (a selective serotonin reuptake inhibitor) is discussed. The reaction is a cyanation, which was previously carried out under Rosemund–von Braun conditions, resulting in a low yield and a difficult purification to purge the copper metal residues. The group looked to improve the process for the final step by using an alternative transition metal catalysis and continuous stop-flow microwave heating. After reaction optimisation of the palladium catalyst, ligand, solvent and additive, the process was taken into batch heating in the microwave. As can be seen in Scheme 4.23, four cycles of microwave heating produced 47 g of citalopram in around 40 min, after a simple filtration work up, to give material of >98% purity. A larger run of 11 cycles produced 150 g in around 2 h. This level of purity and the decrease in production time is a marked improvement on the existing route.

In a paper from Guare *et al.* at Merck,[59] the displacement of a bromopyridine with amines is discussed. The compound is an intermediate *en route* to a series of HIV integrase inhibitor analogues and large amounts of material were required. The process was scaled up in the microwave over 21 batches to yield a total of 415 g in excellent purity following simple filtration, as depicted in Scheme 4.24.

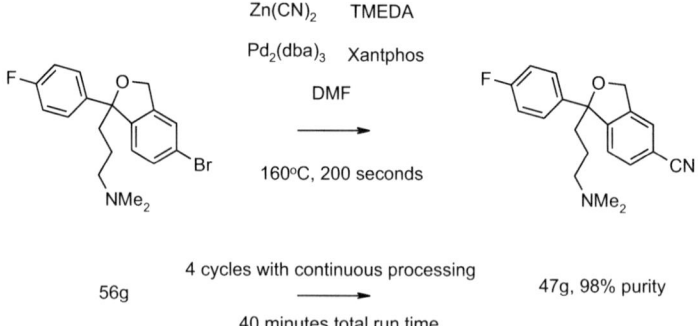

Scheme 4.23 Scale-up of the cyanation step of citalopram under microwave conditions.

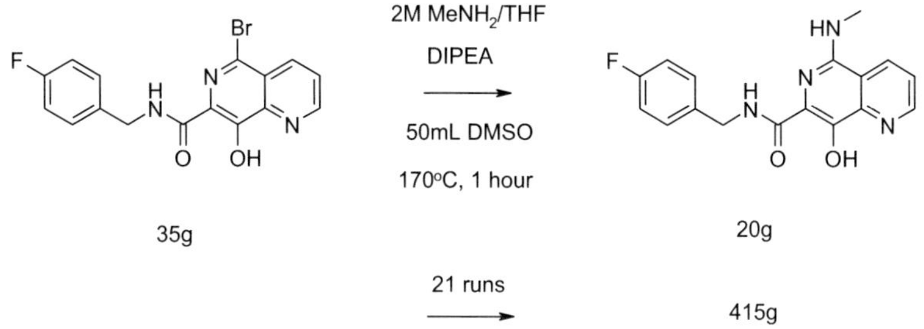

Scheme 4.24   Scale-up of the aromatic amination *en route* to HIV integrase inhibitors.

4.5   Conclusion

It is hoped that in this short review, the reader has been given:

- a sense of the wide range of chemistry that can be undertaken successfully using a microwave reactor;
- an appreciation that many of these reactions can be performed both more quickly and more effectively by using microwave heating; and
- some practical considerations to take into account when attempting to convert standard conditions into microwave conditions.

With these points in mind, it is also hoped that the reader will have no fear of using a microwave reactor to perform their reactions, and will quickly come to appreciate the enormous utility of microwave assisted synthesis.

References

1. R. Gedye, F. Smith, K. Westaway, H. Ali, L. Baldisera, L. Laberge and J. Rousell, *Tetrahedron Lett.*, 1986, **27**, 279–282.
2. R. J. Giguere, T. L. Bray, S. M. Duncan and G. Majetich, *Tetrahedron Lett.*, 1986, **27**, 4945–4958.
3. A. Loupy, *Microwaves in Organic Synthesis*, Wiley-VCH Verlag GmbH & Co, Weinheim, 2002.
4. C. O. Kappe, A. Stadler, *Microwaves in Organic and Medicinal Chemistry*, Wiley-VCH Verlag GmbH & Co, Weinheim, 2005.
5. J. Tierney, P. Lidström, *Microwave Assisted Organic Synthesis*, Blackwell, Oxford, 2005.
6. C. O. Kappe, D. Dallinger and S. Murphree, *Practical Microwave Synthesis for Organic Chemists: Stategies, Instruments, and Protocols*, Wiley-VCH Verlag GmbH & Co, Weinheim, 2009.
7. P. Lidström, J. Tierney, B. Wathey and J. Westman, *Tetrahedron*, 2001, **57**, 9225–9283.
8. C. O. Kappe, *Angew. Chem., Int. Ed.*, 2004, **43**, 6250–6284.
9. C. O. Kappe and D. Dallinger, *Nat. Rev. Drug Discovery*, 2006, **5**, 51–63.

10. C. O. Kappe and D. Dallinger, *Mol. Diversity*, 2009, **13**, 71–193.
11. N. E. Leadbeater, (ed.), *Tetrahedron*, 2006, **62**(19), 4633–4732.
12. http://www.biotagepathfinder.com/texts.jsp?textName = cookbook.
13. M. A. Herrero, J. M. Kremsner and C. O. Kappe, *J. Org. Chem.*, 2008, **73**, 36–47.
14. D. R. Baghurst and D. M. P. Mingos, *Chem. Soc. Rev.*, 1991, **20**, 1–47.
15. C. Gabriel, S. Gabriel, E. H. Grant, B. S. Halstead and D. M. P. Mingos, *Chem. Soc. Rev.*, 1998, **27**, 213–223.
16. M. J. Gronnow, R. J. White, J. H. Clark and D. J. Macquarrie, *Org. Process Res. Dev.*, 2005, **9**, 516–518.
17. http://www.cem.com/.
18. http://www.biotage.com/DynPage.aspx.
19. http://www.milestonesci.com/.
20. K. Olofsson, A. Hallberg and M. Larhed, in *Microwaves in Organic Synthesis*, ed. A. Loupy, Wiley-VCH Verlag GmbH & Co, Weinheim, 2002.
21. J. Hassan, M. Sévignon, C. Gozzi, E. Schultz and M. Lemaire, *Chem. Rev.*, 2002, **102**, 1359–1469.
22. P. J. Hajduk, M. Bures, J. Praestgaard and S. W. Fesik, *J. Med. Chem.*, 2000, **43**, 3443–3447.
23. P. Cao, J. Qu, G. Burton and R. A. Rivero, *J. Org. Chem.*, 2008, **73**, 7204–7208.
24. T. R. Rheault, K. H. Donaldson and M. Cheung, *Tetrahedron Lett.*, 2009, **50**, 1399–1402.
25. P. Walla and C. O. Kappe, *Chem. Commun.*, 2004, 564–565.
26. H. Gold, M. Larhed and P. Nilsson, *Synlett*, 2005, **10**, 1596–1600.
27. G. W. Harbottle, N. Feeder, K. R. Gibson, M. Glossop, G. N. Maw, W. A. Million, F. F. Morel, S. Osborne and C. Poinsard, *Tetrahedron Lett.*, 2007, **48**, 4293–4296.
28. E. Petricci and M. Taddei, *Chim. Ogg.*, 2007, **25**(3), 40–45.
29. J. Salvadori, E. Balducci, S. Zaza, E. Petricci and M. Taddei, *J. Org. Chem.*, 2010, **75**, 1841–1847.
30. E. Petricci, A. Mann, A. Schoenfelder, A. Rota and M. Taddei, *Org. Lett.*, 2006, **8**(17), 3725–3727.
31. K. Leijondahl, A. L. Fransson and J.-E. Bäckvall, *J. Org. Chem.*, 2006, **71**, 8622–8625.
32. P. G. M. Wuts and T. W. Greene, *Protective Groups in Organic Synthesis*, 4th edn, 2007, Wiley-Interscience, Hoboken, N.Y., pp. 725–735.
33. V. H. Rawal and M. P. Cava, *Tetrahedron Lett.*, 1985, **26**(50), 6141–6142.
34. J. Choy, S. Jaime-Figueroa, L. Jiang and P. Wagner, *Synth. Commun.*, 2008, **38**, 3840–3853.
35. V. Åberg, F. Norman, E. Chorell, A. Westermark, A. Olofsson, A. E. Sauer-Eriksson and F. Almqvist, *Org. Biomol. Chem.*, 2005, **3**, 2817–2823.
36. E. Van der Eycken and C. O. Kappe, *Microwave-Assisted Synthesis of Heterocycles, Topics in Heterocyclic Chemistry*, Springer-Verlag, Berlin, Heidelberg, 2006, **1**.
37. W. D. Shipe, F. Yang, Z. Zhao, S. E. Wolkenberg, M. B. Nolt and C. W. Lindsley, *Heterocycles*, 2006, **70**, 655–689.

38. T. Guo, R. C. Hunter, R. Zhanga and W. J. Greenlee, *Tetrahedron Lett.*, 2007, **48**, 613–615.
39. A. Reichelt, J. R. Falsey, R. M. Rzasa, O. R. Thiel, M. M. Achmatowicz, R. D. Larsen and D. Zhang, *Org. Lett.*, 2010, **12**, 792–795.
40. A. Foucourt, C. Dubouilh-Benard, E. Chosson, C. Corbière, C. Buquet, M. Iannelli, B. Leblond, F. Marsais and T. Besson, *Tetrahedron*, 2010, **66**, 4495–4502.
41. S. Fustero, D. Jimenez, M. Sanchez-Rosello and C. del Pozo, *J. Am. Chem. Soc.*, 2007, **129**(21), 6700–6701.
42. M. Jida, S. Malaquin, R. Deprez-Poulain, G. Laconde and B. Deprez, *Tetrahedron Lett.*, 2010, **51**, 5109–5111.
43. H. C. Kolb and K. B. Sharpless, *Drug Discovery Today*, 2003, **8**(24), 1128–1137.
44. C. D. Hein, X-M. Liu and D. Wang, *Pharm. Res.*, 2008, **25**(10), 2216–2230.
45. C. R. Becer, R. Hoogenboom and U. S. Schubert, *Angew. Chem., Int. Ed.*, 2009, **48**, 4900–4908.
46. A. D. Moorhouse and J. E. Moses, *Synlett*, 2008, **14**, 2089–2092.
47. J.-J. Shie and J.-M. Fang, *J. Org. Chem.*, 2007, **72**, 3141–3144.
48. Y. Wang, R. L. Miller, D. R. Sauer and S. W. Djuric, *Org. Lett.*, 2005, **7**(5), 925–928.
49. A. Lew, P. O. Krutzik, M. E. Hart and A. R. Chamberlin, *J. Comb. Chem.*, 2002, **4**(2), 95–105.
50. M. Damm and C. O. Kappe, *J. Comb. Chem.*, 2009, **11**(3), 460–468.
51. C. D. Weaver, D. J. Sheffler, L. M. Lewis, T. M. Bridges, R. Williams, N. T. Nalywajko, J. P. Kennedy, M. M. Mulder, S. Jadhav, L. A. Aldrich, C. K. Jones, J. Marlo, C. M. Niswender, M. M. Mock, F. Zheng, P. J. Conn and C. W. Lindsley, *Curr. Top. Med. Chem.*, 2009, **9**, 1217–1226.
52. L. N. Aldrich, E. P. Lebois, L. M. Lewis, N. T. Nalywajko, C. M. Niswender, C. D. Weaver, P. J. Conn and C. W. Lindsley, *Tetrahedron Lett.*, 2009, **50**, 212–215.
53. A. Stadler and C. O. Kappe, *J. Comb. Chem.*, 2001, **3**, 624–630.
54. H. Lehmann and L. LaVecchia, *Org. Process Res. Dev.*, 2010, **14**, 650–656.
55. A. Stadler, B. H. Yousefi, D. Dallinger, P. Walla, E. Van der Eycken, N. Kaval and C. O. Kappe, *Org. Process Res. Dev.*, 2003, **7**, 707–716.
56. I. R. Baxendale, J. J. Hayward and S. V. Ley, *Comb. Chem. High Throughput Screen.*, 2007, **10**, 802–836.
57. M. Damm, T. N. Glasnov and C. O. Kappe, *Org. Process Res. Dev.*, 2010, **14**, 215–224.
58. M. R. Pitts, P. McCormack and J. Whittall, *Tetrahedron*, 2006, **62**, 4705–4708.
59. J. P. Guare, J. S. Wai, R. P. Gomez, N. J. Anthony, S. M. Jolly, A. R. Cortes, J. P. Vacca, P. J. Felock, K. A. Stillmock, W. A. Schleif, G. Moyer, L. J. Gabryelski, L. Jin, I.-W. Chen, D. J. Hazuda and S. D. Young, *Bioorg. Med. Chem. Lett.*, 2006, **16**(11), 2900–2904.

CHAPTER 5

Continuous Flow Chemistry in Medicinal Chemistry

MARTYN DEAL

Radleys, Shire Hill, Saffron Walden, Essex, CB11 3AZ, UK

5.1 Introduction

Continuous flow chemistry is currently receiving much attention as a new technology capable of enhancing productivity in chemistry laboratories across a diverse range of industries and applications, from pharmaceutical and agrochemical, to petrochemical, and fine and speciality chemicals. The basic principle of running a chemical process in a continuous manner, as opposed to conventional batch-based methods in a flask or tank, is hugely attractive for a variety of reasons.

The potential for infinite scalability, determined by the length of time a process is left running rather than the size of the reactor, is a major driving force. Once a synthesis has been developed and optimised to operate in a continuous flow manner, initially on a small scale where only a few milligrams of material are needed for analytical purposes, no further changes to operating parameters or equipment set-up should be necessary to generate grams, kilograms or even tonnes of material. It has long been recognised in chemical manufacturing that replacing large stirred-tank reactors in expensive dedicated chemical plants with tube or pipe reactors offers enormous benefits in flexibility, space requirements and economics, and it is here that continuous flow chemistry has been most widely employed to date.[1–3]

RSC Drug Discovery Series No. 11
New Synthetic Technologies in Medicinal Chemistry
Edited by Elizabeth Farrant

More recently it has become apparent that extending the concept of continuous flow to as early a stage as possible in the research and development (R&D) process can enhance gains observed in manufacturing. An explosion in the number of publications appearing in the chemical literature over the last year or two describing chemistry performed in continuous flow indicates high levels of interest in both industry and academia. The increasing availability of commercial equipment has allowed technology and capabilities to expand. The development of flow reactors has advanced in a variety of formats and the integration of other common chemistry processes such as purification and analysis have now been demonstrated.

The seamless transfer of chemistry from R&D through to manufacturing is the ultimate goal. Other factors, such as cost of goods and environmental aspects, come into consideration on scaling up, but the principle of developing a synthetic step that can be performed on any scale without needing to change any parameters or apparatus is extremely attractive.

The enhanced performance and efficiency through higher yielding chemistry with fewer side products, created by better control of processes such as reagent addition, mixing and heat transfer; the ability to create reaction conditions not easily achievable in a round-bottomed flask (for example at elevated pressures with super heated solvents); and the inherent safety benefits resulting from only handling small quantities of hazardous reagents and intermediates at any one time are further driving forces behind the broad interest of the community, and will be explored further in this chapter.

5.2 Benefits of Flow Chemistry

The advantages of performing chemistry in a continuous flow manner can be wide reaching but will vary depending on the individual nature of the synthesis under consideration. In some cases there may be very obvious specific benefits, in others it may be an accumulation of several factors or advantages stemming from the process as a whole. However, benefits can generally be broken down into the following categories.

5.2.1 Thermal Control

A successful outcome from a chemical synthesis can often be heavily reliant on the ability to accurately control input or removal of heat from the reaction mixture. The capability to do this quickly and efficiently, allowing consistent and repeatable conditions to be maintained, is a fundamental requirement for all but the most robust chemical processes. Particularly problematic are reactions requiring sub-ambient temperature control, or those generating large amounts of heat, especially at the time of mixing. These problems will generally be magnified as the scale on which a reaction is being performed is increased, resulting in anything from irreproducible chemistry caused by excess reagent degradation or side-product formation, through to unsafe or dangerous working situations.

Working in a batch reactor, typically one reagent will be preloaded into the vessel in a solvent of choice and a second reagent added to it, either as a single dose or in a controlled manner over a period of time, whilst attempting to maintain a consistent temperature for the batch as a whole. For reactions requiring sub-ambient conditions, it is difficult to maintain the reagent being added at the same temperature as that of the material in the vessel, resulting in a temperature differential between the two materials when they meet. For an exothermic reaction, where heat is generated at the point reagents mix, localised temperature gradients will be created within the reactor. If the reaction vessel is small, and stirring efficient, this might not be significant since excess heat can be distributed and removed quickly through the vessel walls by a cooling medium, but as the reaction vessel size increases this is likely to become more problematic.

The design of a continuous flow reactor is ideal for overcoming these issues, resulting in much more precisely controlled temperature conditions. Small diameter, circular channels either embedded in a chip or plate type reactor, or a reactor formed from tubing, provide an optimum reactor-wall surface area to reaction volume ratio, maximising heat transfer to and from the reaction medium. The materials routinely employed in both reactor formats, stainless steel and glass, possess good thermal conductivity, ideal for rapid heat exchange.

Thus heat can readily be applied to a reaction flow stream, generally by means of electrically controlled heating elements either directly embedded in the reactor, or possibly through an independently heated plate to which the reactor is attached. Sub-ambient temperature control is similarly achieved by means of electrically controlled Peltier cooling units, or alternatively through the use of chilled recirculators, which again can either be directly embedded in the reactor or externally mounted.

Careful design of the flow reactor may allow for reagents to be pumped through a pre-mixing temperature-controlled zone, where they can be rapidly heated or cooled to the required temperature before bringing them together. This allows precise thermal control of the mixing process, and uniform and reproducible temperature profiles to be maintained throughout the reactor.

Thermal control of the mixing region is also of great importance for exothermic reactions, as it is here that large amounts of heat are most likely to be generated. This heat is difficult to remove in a controlled way in all but the smallest batch reactors, with the consequences of reagent and product degradation or, in extreme cases, thermal runaway, but can be rapidly dissipated through the large surface area created by the walls of the flow reactor.

5.2.2   Mixing

A batch reactor, either on a lab scale, where stirring is typically achieved through a magnetic stirring bar, or on plant and manufacturing scale, where an

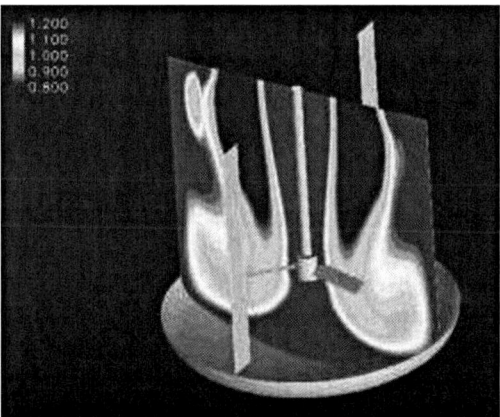

Figure 5.1 Thermal gradients created through inefficient mixing in a batch reactor.

appropriately sized stirrer blade driven by an overhead motor provides the means of mixing the vessel contents, will undoubtedly suffer from non-uniform reagent distribution, particularly at the point and time of reagent addition. Temperature gradients created by localised eddy currents and slow heat transfer between the vessel wall and the centre of the reactor (Figure 5.1) will also almost certainly be present. Neither scenario is conducive to providing uniform and reproducible conditions either within a batch or between batches. Major efforts are made to overcome this on plant and manufacturing scale, often tailored to the individual process through careful design of vessel and stirrer blade or by introduction of baffles into the reactor.

Continuous flow chemistry provides an alternative solution. Here, relatively small quantities of materials are brought together at any one time at the point of mixing, allowing the process to be carefully controlled. Once mixed, either through turbulence or laminar diffusion, depending on the mixing device (see Section 5.4), the reaction mixture is able to pass through the main body of the reactor, where any chemical reaction takes place.

A good mixing performance in a reactor is most important for reactions with fast kinetics or where fast competitive side reactions are possible. The Bourne reaction[4] has been widely used as an example of such a reaction for studying mixing achieved with a range of continuous flow devices, and demonstrating advantages over conventional batch based reactors.[5,6]

Many different strategies have been employed to improve the situation further. Splitting reagent flow streams into many tiny channels before the point of mixing encourages laminar diffusion type mixing, whilst also improving heat transfer through the increased surface area. Alternatively, varying the shape and dimensions of the mixing zone and the introduction of additional static mixers encourage better turbulent behaviour. In both cases, excellent uniformity of reagent distribution can be achieved reliably and repeatably, alongside the high degree of thermal control.

5.2.3 Stoichiometric Control and Selectivity

The manner of reagent addition in a flow reactor is ideal for accurately controlling the relative stoichiometric ratio. As has already been mentioned,
typically a batch reactor will be preloaded with one reactant, and a second
reactant added to it either in bulk or portionwise over a period of time. This
may not present a problem for chemistry with slow kinetics, where reagents
have time to mix and distribute themselves evenly within the reaction medium
before reaction occurs. However, for fast chemistry, temporary localised high
concentrations of one reactant over the other may give rise to stoichiometric
inbalances leading to unwanted side-product formation. For example, the
reaction of a diamine with an acylating agent is likely to suffer from bis-acy-
lation if localised high concentrations of acylating species are present (Scheme
5.1). However, if the stoichiometry could be accurately controlled such that the
ratio of the two reactants was always 1 : 1, single addition at the most active
amine would always occur. This concept could lead to improved selectivity or
possibly avoid the use of protecting groups in some reactions.[7–9]

   Additionally, it is possible to isolate the product from the reactants in a flow
reactor in a way that is impossible to achieve in a batch reactor, reducing the
risk of secondary reactions and side-product formation. Once the product has
formed it is automatically isolated by the flow of the system from the source of
fresh reagent addition, and it therefore becomes impossible for the two to meet.
A batch reactor will always have product molecules in close contact with
reactants until the point when all reactants have been consumed, leaving the
potential for secondary reactions to occur.

5.2.4 Safety

Safety is obviously of paramount importance when considering undertaking
any chemical process and, once again, the nature of performing a chemical
synthesis in a continuous manner offers advantages. Necessarily, for a batch

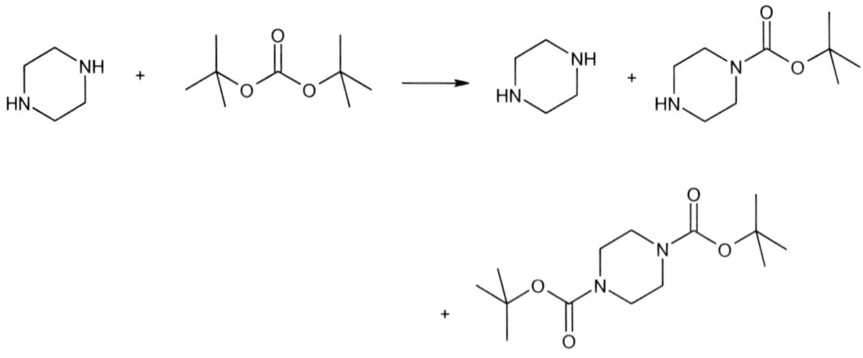

Scheme 5.1   Formation of unwanted bis-acylated side product.

process, all materials required for the synthesis, along with products formed, will be present in the reactor for the duration of the synthesis. Any adverse effect leading to a hazardous situation during this process, for example the breakage of the reaction vessel or a lack of thermal control leading to thermal runaway, will therefore impose itself on the whole contents of the reactor. Associated safety implications will be directly proportional to the scale that the process is being performed on.

Conversely, for a process carried out in a continuous reactor, risks can be significantly reduced. Firstly, through the ability for better process control already described, (thermal, mixing *etc.*) hazardous situations are less likely to develop. Secondly, the amount of material being handled at any one time will be only a small fraction of the total process amount. The majority of materials will be present in a relatively safe neutral and isolatable environment, either as unreacted reagents waiting to be processed, or as product collected on exiting the reactor. The impact of any adverse event, most likely to occur within the reactor (or possibly supply lines to and from it), will therefore be minimised, and no longer proportional to the scale of the process as a whole.

A secondary but not insignificant benefit arising from handling only a fraction of the material at any one time is of more economic consideration. Loss from a batch process, either through a safety related event, or failure of the chemistry from any number of possible causes, will naturally impact on the whole batch. This could have severe financial implications, depending on the nature of the materials, and again will be directly proportional to the scale. A continuous process offers the potential for minimising this impact but, for anything other than the most obvious cases, relies on the availability of some form of regular system monitoring. Sampling and analysis of the output from the reactor, or possibly some form of in-line continuous analytics would allow the process to be carefully monitored from start to finish. At the first sign of any problem, supply of reagents to the reactor can be stopped, leaving already processed and collected product, together with unprocessed reagents, unharmed. The reactor can then be isolated from the system and any damaged materials removed. In this way, loss due to process failure can be limited to one reactor volume, generally a tiny fraction of the total material.

5.2.5 Enhanced Process Parameters

Possibly one of the most attractive prospects of operating in a continuous system from a chemist's point of view is the opportunity to advance the boundaries of chemistry space created by the availability of extended processing conditions. The development and application of microwave enhanced synthesis[10] has featured increasingly widely in the armoury of the synthetic organic chemist over the past decade or so. The ability to readily perform synthesis at elevated pressures, under relatively safe conditions, has proved advantageous. The added attraction of working with superheated solvents

Table 5.1 Boiling points of common solvents at normal and elevated pressure.

Solvent	Boiling point (°C) at 1 bar	Boiling point (°C) at 3 bar
Dichloromethane	40	62
Methanol	65	91
Tetrahydrofuran	66	92
Ethyl Acetate	77	105
Toluene	110	144
Dimethylformamide	153	195

above their normal boiling points (Table 5.1) allows chemistry to be performed under conditions not previously available and opens up scope to drive reactions down different pathways and develop "new" chemistry.[11]

The one limiting factor identified in applying microwave chemistry widely is the restriction on the operating scale, partially stemming from the design of the instrumentation itself, but more significantly from the additional safety considerations attached to increasing the size of the reactor vessel. The technology behind the common laboratory microwave systems is based on monomode cavities, which enable high energy input to be applied to the reactor. However, this forces constraints on the maximum size of cavity, and therefore reactor. Additionally, limited penetration depth of microwaves into the reaction mixture of a few centimetres means that a large scale reaction vessel is unlikely to be effective.

Automated sequential batch processing can be used to increase the scale of chemistry that can be performed in this way, but manufacturing scale quantities are unlikely to be achieved.

There is much debate as to the reasons for the rate enhancement seen in microwave chemistry. Common thought is that it is purely down to the high pressures that can be realised within the sealed microwave reactor vessel, coupled with the ability to utilise superheated solvents (see Chapter 4). Conveniently, continuous flow reactors demonstrate the same ability to operate under such conditions. Operating pressures up to 100 bar can safely be obtained in the majority of commercial flow reactors through the simple addition of a controlled pressure regulator at the end of the reactor flow path. Operating at such pressures does put more exacting requirements on the pumping system to maintain accurately controlled flow rates, but there is a wide range of commercially available pumps capable of operating in such environments. In this way, it is therefore possible to recreate the beneficial conditions utilised in a microwave cavity, generating enhanced reaction kinetics and possibilities for developing new chemistries. The advantage is that it is being done in a system without limitations of working scale and material throughput, which could prove hugely beneficial.

An intriguing extension to this subject is the question of whether the two technologies of microwave and continuous flow chemistry can be married together in a way to give complementary or additive benefits. This will be covered in more depth later (see Section 5.9.1).

5.2.6 Hazardous Reactions

The general safety benefits of operating in continuous flow have already been discussed (see Section 5.2.4), along with the improvements in handling hazardous exothermic processes through better thermal control. There are numerous examples of other reaction classes that benefit from being performed in this way. In general, advantages stem either from the fact that only small amounts of the particular hazardous reagent are being handled, or from situations where the hazardous material is generated *in situ*, and consumed in a short period of time within the reactor.

In many cases, chemists have avoided these reaction classes in recent times because the safety risks involved in performing them are considered too high. Good examples are reactions involving unstable nitrogen compounds, for example azides, diazo species and diazomethane, and those involving hazardous gases such as hydrogen, carbon monoxide and ozone, particularly for applications requiring elevated pressure.

Possibly the most reported and exploited use of flow chemistry in a research situation to date is that of hydrogenation, primarily brought about by the development of equipment specifically to perform this task.[12–14] This utilises the safe, *in situ* generation of hydrogen gas within the system (from electrolysis of water) at a point where it can be simply introduced into the reactant flow line. High pressure applied to the flow stream enables the gas to be compressed and rapidly mixed with reactant immediately prior to entering the reactor, in this case a pre-packed heated cartridge of activated catalyst. The additional benefit of being able to safely contain and handle rarely used highly pyrophoric catalytic materials in this reactor format opens up possibilities for new chemistry applications. Catalyst screening can be easily and safely performed with a wide range of pre-packed activated catalysts, for example RANEY® nickel, rhodium, cobalt and platinum species, at higher pressures and temperatures than would normally be considered. Once an optimum catalyst has been identified, reaction scale-up can be achieved without increasing the amount of catalyst required by simply flowing the reactant–dissolved hydrogen mixture through the catalyst bed for as long as is required. A further interesting consideration is that the dynamics of performing hydrogenation changes between batch mode, where reactants are present in large excess relative to the catalyst, and flow mode, where the catalyst is present in much higher proportion, and often in excess. This in part explains the extremely rapid reaction kinetics often observed, and opens possibilities for cleaner reaction profiles and enhanced selectivities between the reducible functional groups present.

5.2.7 Multistep Processes

One aim when designing a chemical route to a target molecule, be it on a research or manufacturing scale, is always to simplify the overall process as much as possible. The route will invariably involve a multistep chemical

Scheme 5.2   Typical multistep synthesis.

synthesis (Scheme 5.2), with purification and isolation of the chemical inter-mediate from each synthetic step.

In a traditional batch process, this generally involves a discrete operation for each of the synthetic and purification steps, with the whole batch being handled for each operation. It may be possible to streamline operations by performing sequential reactions in the same reactor without intermediate isolation if complete conversion can be achieved. If this is not realised, risk of unreacted reagents and side products interfering with subsequent reactions and the rapid build-up of impurities that occurs with each step place limits on what is feasible.

Flow chemistry opens up some interesting possibilities to capitalise on advances in this area. Cleaner chemistry, for the reasons already highlighted, makes joining two synthetic steps of a sequence together more achievable. Practically, it is a relatively straightforward task to combine the outlet flow from a continuous reactor containing product from the first reaction with another reagent stream to form the inlet flow to a second reactor. However, further improvements can be achieved with the addition of in-line clean-up and pur-ification techniques between the two reactors. Solid-supported scavenger reagents have been much employed as a method of removing impurities from a reaction mixture in recent years[15] and these are ideally suited for utilisation in a continuous flow manner. Supported scavengers can be loaded into cartridges and, with the appropriate fittings, plumbed into the reaction flow line. Immo-bilisation in this way allows them to be positioned and retained in the reaction flow path at the exact point that they are required to remove unwanted materials.

The ultimate goal, using this philosophy, is to create complete continuous synthetic routes to target molecules, where reagents are fed in at appropriate points along the process line and pure product exits at the end of the reactor sequence. Although much is still to be learnt before this can be widely applied, there are already elegant examples published where this has been demon-strated.[16,17] One limiting factor is the finite scavenging capacity of the sca-venger reagent, which must be periodically replaced or regenerated. Manually swapping cartridges when required is an option, but sophistication can be brought to the process with the use of automated flow switching devices that allow multiple cartridges to be set up and used sequentially. However, realis-tically, to achieve large-scale throughput in this multistep manner, purification between steps needs to be attained without the need to intervene in this way.

Efforts to achieve multistep sequences on large scale processes are more focussed on integrating liquid–liquid extraction, distillation and crystallisation techniques into the continuous flow format, and although there are currently no published examples of complete continuous synthetic routes being per-formed on a production scale, it is predicted that this will be realised in the near future.[18,19]

5.3 Limitations and Technology Hurdles

As with the introduction of any new technology, not everything is always as straightforward as would be liked, and there are issues that currently limit capabilities. However, as the technology matures and evolves it is likely that many of these will gradually be resolved.

5.3.1 Solubility of Reagents

A lack of solubility of both reactants and reagents is generally considered to be the primary hurdle for any continuous flow reaction process, as the basic concept relies on the ability to pump liquids of known concentration. Pumping technologies are generally not well suited to handling solids and tend to clog and block if particulate matter is present. Also, depending on the design of the reactor, narrow channels and restriction points within the flow path will be prone to partial or complete blockage. A partial blockage would not necessarily stop the process but could affect process parameters, particularly flow rate, or create a hold up of material in the reactor, leading to inconsistent performance and poor results. A total blockage would prove more problematic, and inevitably lead to a shut down of the system, and possibly to the hazardous situation of internal pressure build up. Although most systems are capable of handling moderate amounts of pressure, this is generally to be avoided. Some pumping technologies are better suited to handling particulates, and can be employed for such purposes. Care must be taken to ensure that reagents added as suspensions are adequately stirred prior to entering the reactor to maintain some degree of consistent composition.

An additional concern with total blockage in a system is how to recover the situation, particularly within a fully sealed and enclosed assembly. It can often be difficult to locate the exact blockage point and not easy to access the site once identified. The introduction of a solubilising agent is seldom possible; the application of ultrasound can sometimes be beneficial in breaking up any accumulation of solids. Usually the simplest solution is to completely remove the blocked component and replace it with a new one. At this point, system modularity and cheap reactor components demonstrate a significant advantage.

5.3.2 Solubility of Products

Less generally considered, but of equal importance, is the solubility of the products and side products formed in the reaction. Although all reagents may be completely soluble entering the reactor careful consideration should be given to the nature of the products formed before beginning a new flow process. Particular care should be taken when the reactor is being operated at elevated temperatures, where there is a tendency for product stream exiting the reactor to be cooled rapidly to ambient temperature resulting in precipitation or crystallisation. Ideally, the flow path should continue to be heated post-reactor and the product stream not cooled until it is collected in an open vessel.

Preferably a flow system should be capable of detecting the first signs of blockage, often indicated by a slight rise in system pressure, and react to remedy the situation, or at least alert the operator with a warning. However, generally the best way to avoid the problem is by careful pre-screening of reagents and products, to identify solvent systems likely to minimise the formation of any solid material.

An alternative strategy particularly suited for reagents or catalysts that do not have good solubility in a required reaction solvent is to employ the material in its solid form packed into a cartridge that can be physically plumbed in to the flow line. Further details and examples of this are described in Section 5.7.

5.3.3   Scale

The challenge of being able to apply flow chemistry from early stage drug discovery on a miniaturised scale to identify a candidate molecule, through to a continuous flow manufacturing process by simply scaling up a synthetic route to the appropriate level, on the same equipment, is still a little way off. Microreactors developed for producing the small amounts of material needed for early stage research are typically capable of producing a few grams of material a day at best, so even running around the clock would result in annual outputs in the range of 1 kg.

5.3.3.1   Scale Out

The concept of "scale out", where a single microreactor is replicated many times so that the same process can be performed in parallel, can be successful in increasing output but suffers from some practical difficulties. Attempting to run parallel reactors from a single pump by splitting the flow equally to each reactor requires not only precisely engineered splitting devices, but also equal performance to be maintained over the full length of all channels. Any deviation in behaviour between channels will produce inconsistent results. Most difficult to maintain are system pressure and flow rate. Any channel generating a slight pressure change (for example from a partial blockage) will cease to flow at the same rate, and the resulting chemistry will experience higher retention, and therefore reaction times, than other channels. The alternative is to provide each channel with its own pumping mechanism, which then requires high levels of reproducibility between pumps, and introduces significant financial implications. If "scale out" is to be used for increasing throughputs, it is likely to be constrained by these factors.

5.3.3.2   Scale Up

More widely employed methods for scaling up currently tend to involve changing the reactor format to accommodate the required increases in material throughput. This can involve increasing the reactor volume, typically by increasing both the length and cross sectional area of the flow channel, or by moving to an alternative complementary flow technology, for example continuous stirred tank or oscillating baffle reactors.[20,21] This obviously introduces a step change in technologies, with consequent changes in operating parameters for factors such as mixing and thermal control. However, often it has been demonstrated that processes are capable of being directly transferred from a small to larger scale reactor format without suffering any ill effects.

A good example of this has been described for the synthesis of a Grignard reagent, pentafluorophenyl magnesium bromide (Scheme 5.3), initially on a laboratory scale, through to continuous plant production.[22] This rapid, exothermic and difficult-to-control reaction, requiring slow reagent addition in batch mode, was identified as an ideal candidate for production in a continuous flow reactor.

Initial studies in a microreactor on a milligram scale showed that residence times of a few minutes were required for complete reaction, and the nature of the mixer and reactor diameter were critical in achieving good results. Scaling the findings to an intermediate scale reactor produced similar results and once again the nature of the mixer was identified as a critical issue. Finally, the methodology was transferred to a pilot plant capable of continuous operation over 24 h, where it was determined that residence times could be as short as five seconds, allowing throughputs of 14.7 kg of product per day.

A further example involved the preparation of an important intermediate in the manufacture of a pharmacologically active molecule,[23] where an existing but problematic batch reaction was converted to a continuous flow process. The first two steps of a three-step synthesis required the oxidation of a secondary alcohol with NMO–TPAP (*N*-methylmorpholine-*N*-oxide–tetrapropylammonium perruthenate) followed by an exothermic and unpredictable trifluoromethylation (Scheme 5.4). Here, reaction optimisation was performed on a millgram scale in a plug flow reactor, but to achieve the required throughput conditions were transferred smoothly to a non-plug flow reactor.

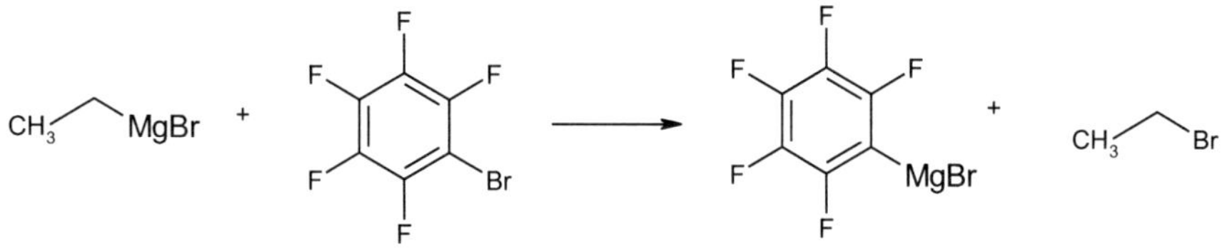

Scheme 5.3 Scale up of a Grignard reagent synthesis to 14.7 kg product per day.

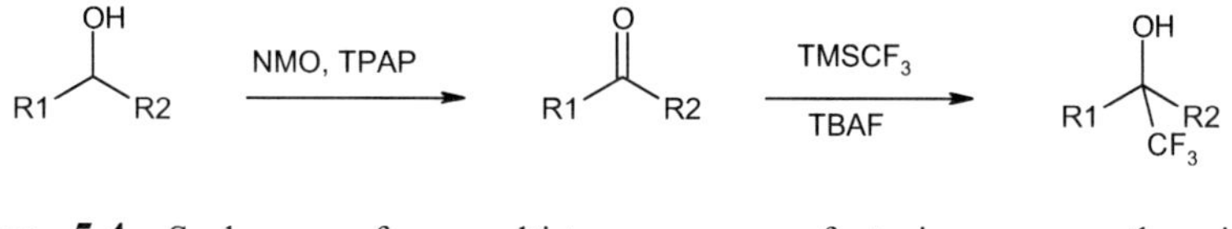

Scheme 5.4 Scale up of a multistep sequence featuring an exothermic trifluoromethylation reaction.

A further important point is that often there are wider implications to consider when scaling up from research to manufacturing scale. Factors such as cost of goods and environmental issues, such as waste disposal and atom efficiency, are given little consideration at the research scale but become hugely important for manufacturing. Thus in reality, at some point during the development process it is highly likely that a change in reagents and/or chemical route will always be required, which will provide a perfect opportunity to change the reactor format.

5.3.4 Other Issues

Highly corrosive reagents, for example strong acids, can quickly cause a problem if inappropriate equipment has been utilised. To avoid such situations, every component in direct contact with the flow stream must be manufactured from a fully chemically compatible material.

The generation of gaseous side products in a reaction is common, and of little consequence in a batch reactor where they generally have a means of controlled release. However, the formation of gases within a flow reactor can have much more serious consequences, particularly if the evolution is vigorous. A build-up of gas bubbles will occur within the reaction fluid, and the high volume of gas produced relative to liquid exacerbates the situation. The space required to accommodate these bubbles results in reaction solution being propelled through the reactor faster than anticipated, giving unreliable residence times at best, and violent emptying of the reactor in more extreme cases. A simple solution exists, as employing a back pressure regulator on the outlet of the reactor usually resolves the situation. Pressurising the reactor in this way keeps any gas generated compressed to minimum volume, and avoids disrupting flow rates and reaction time.

The accuracy of flow control through the reactor is important for two reasons. Firstly, the overall flow rate of the system is directly related to the reaction time, so it is important to maintain it at a consistent level particularly where time-sensitive reactions are being performed. Secondly, if reagents are being pumped into the reactor *via* their own individual pumps it is important that the relative flow rates of the pumps are balanced and remain consistent. Failure to achieve this may make it difficult to maintain reliable stoichiometric ratios of reagents, which would have serious consequences for the quality and reproducibility of the reaction product. Maintaining pumps in good working order, with careful calibration of flow rates at regular intervals, helps address the problem.

5.4 Flow Dynamics: Some Basic Theory

5.4.1 Reynolds Number

The flow properties of a fluid in a channel can be described in mathematical terms as the ratio of its inertial forces to its viscous forces, and are expressed in terms of its Reynolds number, Re, which is dimensionless, such that:

$$\text{Re} = \rho V D / \mu$$

where ρ = density, V = average velocity in the channel, D = channel diameter, and μ = viscosity.

For low Reynolds numbers below 2000, flow is described as laminar; for a Reynolds number above 4000 flow is considered turbulent. For values in between there may be a mixture of flow characteristics present, and other factors such as the channel surface roughness and flow uniformity become relevant.

So, for example, to achieve laminar flow, *i.e.* flow with a low Reynolds number, a low velocity and small channel diameter combined with a low-density, high-viscosity liquid is optimal.

5.4.2 Laminar Flow

Laminar flow produces parallel, linear individual layers of fluid within the channel at the point where two flow streams come together, which remain in this state with no disruption of layers over the whole length of the channel (Figure 5.2). Hence there is no direct mixing of the two fluids, although diffusion can occur across the boundary layer from one stream to the other. In the case where the two flow streams are immiscible liquids, partitioning of components dissolved in one or other phase can take place by diffusion across the boundary layer.

5.4.3 Turbulent Flow

Turbulent flow is characterised by disorderly patterns of flow within the channel, with eddy currents, random fluctuations and streamlines intertwining, although the overall flow remains in one direction. This results in a chaotic and non-reproducible mixing of the components from two flow streams combining together (Figure 5.3). The efficiency of mixing is determined by such parameters

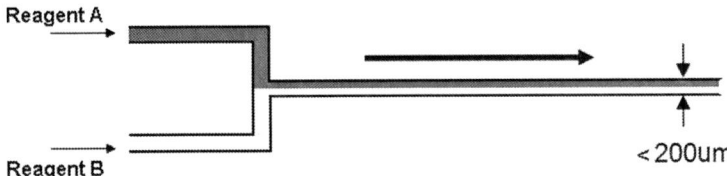

Figure 5.2 Laminar flow from two flow streams.

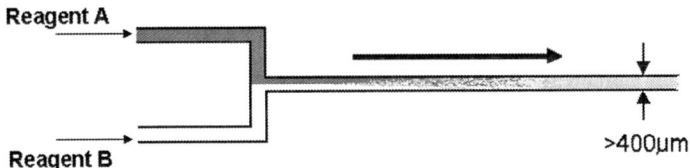

Figure 5.3 Turbulent flow from two flow streams.

as fluid flow rate and the shape and dimensions of the cross section of the channel. Turbulent flow can be encouraged or enhanced by the introduction of features, termed static mixers, within the channel. Such static mixers, typically baffles or helical coils, develop shearing and blending motions within the fluid flow, which promote mixing.

5.4.4 Parabolic and Plug Flow

Pressure-driven flow through a channel produces a parabolic velocity profile in the fluid, whereby the central core of the fluid moves faster through the system than the outer layers. Figure 5.4 shows a cross section of such a channel: forces at the fluid/wall interface introduce a drag effect that restricts flow. Theoretically, under true laminar flow conditions, the flow velocity at the fluid boundary layer next to the channel wall is zero, and at the centre of the channel is twice the average bulk velocity. Under conditions where some degree of turbulent flow exists, the effect of radial diffusion within the channel creates movement at the wall interface and slows the central channel.

Where conduction of heat is occurring between the fluid and channel walls, parabolic flow can give rise to temperature gradients within the system, the faster moving central channel being less able to transfer heat than the interface layer.

In practical terms the implications of parabolic flow are that different components within the system can experience different conditions. The length of time each remain in the reactor and the parameters (*e.g.* temperature) they are exposed to during that time will vary, and it should be remembered that quoted experimental conditions, for a chemical synthesis for example, are generally referring to average conditions.

Further consequences may impact on a plug flow process, whereby a "plug" of material of defined volume is inserted into a channel (*via* an injection loop for example) containing an inert carrier fluid moving at a given velocity. Parabolic flow will distort the plug from its original shape and volume, creating a much larger plug as it travels down the channel. The front and back end diffuse into the carrier fluid, causing the contents of the plug to dilute as it does so, creating a concentration gradient along the plug. If attempting to collect the plug as it exits the channel based on a theoretical calculated time, careful consideration should be given to this diffusion effect.

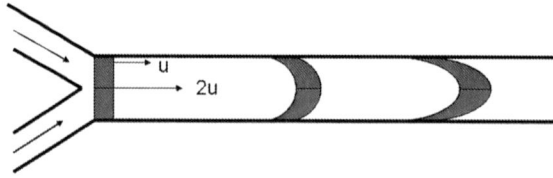

Figure 5.4   Parabolic flow in a reaction plug formed from two reagent inputs.

5.4.5 Residence Time Distribution

The passage of a liquid through a length of tubing is determined in its simplest form by parameters such as its flow velocity and the dimensions (length, cross sectional area) of the flow path. The residence time of a continuous flow reactor is defined as the time that any material stays within the reactor. It is generally referred to as an average figure calculated by:

$$T = V/u$$

where T = residence time (min), u = flow velocity (ml min^{-1}), and V = internal reactor volume (ml).

In an ideal situation, all components are flowing at identical rates and therefore each component will have an identical residence time. However, the range of flow rates generated within the channel through the phenomenon of parabolic flow leads to an equivalent range of residence times, and the distribution of these is referred to as the residence time distribution of the system.[24,25]

In the situation where chemical synthesis is being performed in the flow stream, this residence time distribution may not be important in situations where reaction time is not critical. In cases where, for example, unstable intermediates are being formed and quickly quenched by a component entering the flow stream at a defined point (and therefore time) along the flow path, then residence time distribution may take on more significance. Side-product formation is more likely to result from components deviating from the ideal residence time.

5.4.6 Electroosmotic Flow

The phenomenon of electroosmotic flow (EOF) offers interesting potential as a further differentiating feature between batch and flow chemistry. Glass chips or fused silica tubing employed in microfluidic reactors comprise channels lined with SiOH groups, which in appropriate conditions (at neutral or alkaline pH) create a negatively charged surface to the channel. Any positively charged materials present in the solution contained within the channel will be attracted to the charged surface, creating a charged liquid boundary layer (Figure 5.5).

Applying a high electric potential between the ends of the channel results in the movement of the boundary layer, with positively charged counter ions moving towards the cathode. This movement quickly transfers to the bulk of the material in the channel through viscous forces, resulting in a general movement of the material, known as electroosmotic flow (Figure 5.6).

The ability to move solvents and reagents in this way can create situations not possible in a batch reactor, offering an opportunity for the development of "new chemistry". An example of this[26] shows improved yields and a reversal of Z/E selectivity in a series of Wittig reactions performed under an EOF pumping regime. Further examples can be found covering a range of different chemistries.[27–29] However, the exacting preparation required to achieve the necessary

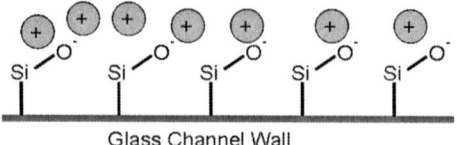

Figure 5.5 Charged liquid boundary layer from a treated glass channel wall.

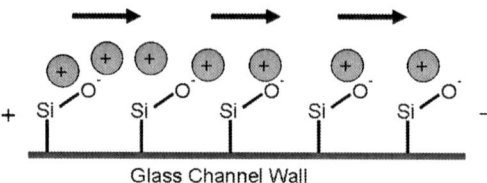

Figure 5.6 Electroosmotic flow on applying an electric potential.

internal surface characteristics of the channel, along with the often unpredictable behaviour of solvents and reagents to the applied electrical potential, means the technique has yet to be fully exploited, and makes it unlikely to be of generic application.

5.5 Experimental Set-up

An increasing number of commercial flow chemistry systems are becoming available, offering the chemist a choice of scale, complexity, capability and price. Each has its own unique features and all are generally designed in such a way that a system can be built up and added to in a modular way allowing for expansion as the user's needs develop. Undoubtedly, hardware development will continue to extend the range of capabilities in the future, so it is important that systems are flexible enough to be upgraded in this way without the expense of having to replace the whole set-up.

  Although commercial systems vary considerably in what they offer, there are some key components that are common to all.

5.5.1 Pumps

An essential component of any flow chemistry arrangement is a pumping system capable of operating accurately over a range of flow rates. The most basic system requires only a single pump to control the main flow stream in the reactor. This requires manual addition of reagents into the front end of the reactor and is likely to be limiting in its capabilities. Much more control can be achieved with individual pumps delivering each of the reagents to the reactor. The total flow rate of the system then becomes the sum of the individual reagent

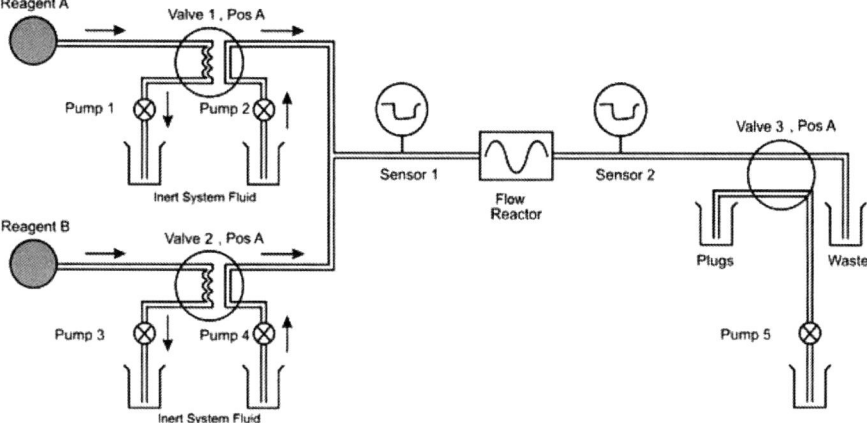

Figure 5.7   Reagent in "load" position for a sample loop.

flow rates. This also allows the possibility of varying the reagent stoichiometry by adjusting the ratio of the individual reagent flow rates, an important feature whilst reaction parameters are being optimised.

In situations where reagents come into direct contact with the pumping mechanism, attention must be paid to the chemical compatibilty of the internal components of the pump heads. Strategies avoiding direct chemical contact with the workings of the pump exist, whereby reagents are introduced into a flowing stream of pure solvent downstream from the pump. Reagents can be loaded into a sample loop, either manually with a syringe, or through an automated liquid handling device. Switching sample loops from the "load" position (Figure 5.7) to the "inject" position (Figure 5.8) delivers reagents into the flow stream.

This set-up is most appropriate for creating "plug flow", where a plug of reaction mixture of defined volume can be formed and passed through the reactor, sandwiched between sections of an inert carrier solvent. It is more difficult to operate in this mode when a continuous stream of reaction mixture is required.

An alternative is to use peristaltic-type pumping systems which operate through indirect contact with the reaction medium. These systems tend to be less accurate, particularly at very low flow rates, and are not suitable for applications where the system is pressurised.

5.5.2   Mixing

One of the major benefits of flow chemistry is the accurate control and reproducibility introduced for the mixing of reagents at the start of the process line. As has been covered earlier, this presents opportunities for faster, cleaner chemistry and enables very good control particularly of exothermic reactions, where heat generated at the point of reagent mixing can be rapidly removed. It should be remembered that mixing performance is very much governed by the

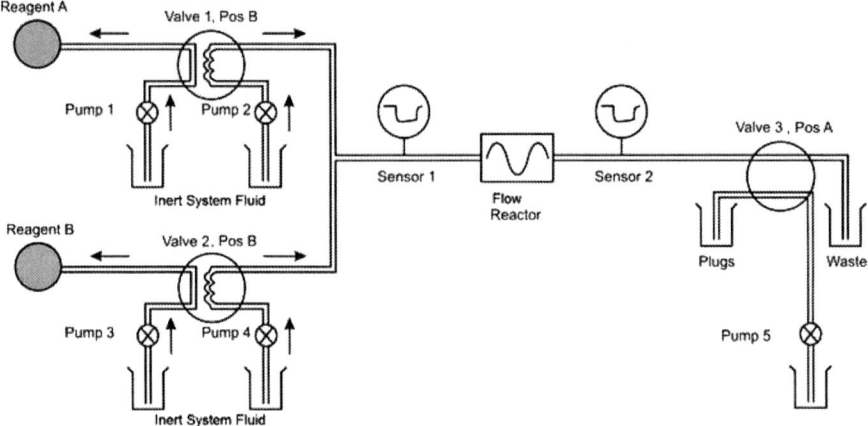

Figure 5.8   Reagent in "inject" position for the reactor.

nature of the mixer being used. For microfluidic processes operating in narrow channels, mixing can be achieved solely by diffusion across laminar flow boundary layers. In systems with larger channel diameters, turbulent flow produces the mixing either with or without the aid of static mixing or baffle devices. It is possible to design sophisticated mixing devices to maximise performance through increasing contact surface areas, depending on the nature of the particular piece of chemistry. However, not all reactions require such exacting set ups, and in many cases a very simple T-piece connector will suffice for the purpose of bringing individual reagent streams together.

5.5.3   The Reactor

The reactor, where all the chemical reaction is designed to take place, sits at the heart of every flow synthesis set-up. Reagents can be introduced through separate channels into the front end, where they meet, mix and start to react, or in some cases may be pre-mixed in a separate mixer unit situated immediately prior to the reactor.

Two different formats of reactor are commonly employed, either chip-based or tube-based.

5.5.3.1   Chip Reactors

Chip-based reactors are generally the favoured format for microfluidic systems, where flow rates and working volumes are low. Chips are typically manufactured from high quality glass, offering high chemical compatibility (although silicon, quartz, metals and a range of polymers have also been used). Precise flow channels can be etched into thin sheets of glass using fabrication techniques such as laser microforming, photolithography, powder blasting, and hot embossing. Welding or bonding sheets together produces precision flow paths, which may be tailored to specific requirements. Much attention has been paid to the development of sophisticated attachment ports on the chip surface to the inlet and outlets of channels. Quick and easy assembly and removal of the chip from the flow path is desirable, with minimal dead volumes and no leaks. Chips can be tailor-made for dedicated applications and provide an ideal way to interface chemical synthesis with other processes, in line with the "lab on a chip" concept. Here, the aim is to bring chemistry, analysis, purification and biological screening together in a single chip or a series of connected chips. Working volumes ranging from nanolitres up to a few millilitres can be achieved in a chip format.

The high degree of precision and complexity available inevitably leads to cost implications, with current manufacturing typically customised for a specific application in low numbers. More generic chips capable of being mass-produced do provide a more cost-effective option.

5.5.3.2 Tube Reactors

By far the simplest way to run a flow chemistry process is with the use of a tube-based reactor. A simple length of tubing of the type typically employed in HPLC applications, widely available in a range of diameters and chemically compatible materials, is all that is needed to get started. A large range of available connectors and fittings enables the tube reactor to be connected to peripherals such as pumping systems, reagent injection devices and sample collectors, and provides scope for great versatility. The tube reactor can be wound around heat transfer devices to provide the necessary thermal control and, with a pressure regulator connected in-line, reactions can be run at elevated temperatures and pressures in a relatively safe manner. Careful attention must obviously be paid to the nature and condition of the tubing material and parameters such as its wall thickness if such forcing conditions are being applied. Operating pressures in excess of 100 bar can be readily achieved, for example, with a suitable grade of stainless steel tubing. High operating pressures also require any connectors and attachments in the flow path to be capable of working under such conditions.

5.5.4 Sample Collection

Having passed through the reactor the exiting flow stream containing reaction products needs to be collected in a suitable vessel. Requirements for this will vary depending on the nature of the process being performed. If a true continuous process is in operation, with a constant stream of identical material being produced, nothing more complex than a container large enough to hold the total amount of material is required. However, if discrete plugs of reaction mixtures are being processed, with each plug containing different materials, separated by, for example, an inert solvent barrier, something more sophisticated is required. An automated fraction collector provides the best means of collecting individual plugs in separate containers as well as separating plugs from inert spacing solvent.

Two common strategies are employed for performing this separation. A simple timing process whereby the system can activate the fraction collector according to a calculated theoretical time for each plug to be exiting the reactor, based on parameters of flow rate and internal system volume. The alternative method, potentially offering a higher degree of reliability, requires the use of a suitable detector immediately before the switching valve of the fraction collector. The detector, able to differentiate between a reaction plug and inert spacer, is typically UV based, but other sensors may be appropriate. Continuous monitoring of the flow line allows the collection valve to be switched between waste and collection as determined by the output from the detector. Allowances can be made in both methods for collecting all or parts of the plug such that, for example, concentration changes caused by diffusion at the beginning and end of reaction plugs can be discarded. When only an analytical sample is required a small amount of material can be extracted from the centre of a plug, where conditions are most likely to be representative.

It is possible to bring further sophistication by combining both fraction collection and analysis in one automated operation. Use of a diverter valve allows isolation of a small fraction from the reaction plug to a sample loop forming the input to an analytical device. In this way, samples can be produced, collected, and fully characterised in a completely automated manner.

5.6   Converting a Process from Batch to Flow

In an ideal world, when faced with the prospect of developing a new flow synthesis to a target molecule, thought processes would begin with meeting the requirements necessary to run in flow mode. These can be fundamentally different to the way a conventional batch synthesis would be planned. Important factors such as solubility of reactants and reagents (as well as intermediates, products and side products), maximising reaction concentration and minimising reaction time come to the forefront. In reality, the extensive literature already published and available, which documents synthetic routes developed and optimised for batch-based synthesis, makes the batch conditions difficult to ignore as a starting point.

Are all chemistries going to be suitable to run in a continuous flow manner? The answer is almost certainly no, and some of the technology hurdles and limitations have been highlighted elsewhere. However, it has been estimated that 50% of reactions undertaken in the fine chemical/pharmaceutical industry could benefit from being performed in this way.[30] It is important, therefore, to have an understanding of what makes a reaction particularly suited to a flow environment.

Fast, highly energetic reactions are ideal, where short residence times in the reactor enable a rapid throughput of material (particularly important when scale-up to large quantities is required). Similarly, highly exothermic reactions or reactions limited by heat transfer rates will be enhanced by the excellent thermal transfer and control properties of the reactor, producing safer operating conditions and the likelihood of cleaner chemical conversion.

Reactions involving hazardous or unstable intermediates benefit from the fact that, compared to a batch process, only limited amounts of such intermediates are produced and present at any one time. These have the opportunity to form stable products as soon as they have been formed in an environment where they are physically isolated from unreacted reactants. The contained nature of the flow reactor means that there is minimal chance of user exposure to any such hazardous intermediate before it is consumed.

Reactions requiring careful process control to achieve high selectivity or chiral purity will similarly benefit from the unique properties that flow reactors demonstrate in being able to effectively segregate reactants from products, minimizing the risk of over-reaction.

High pressure reactions are particularly difficult to perform safely in a batch process, and are thus generally under-utilised or completely avoided in both research and manufacturing. Flow reactors offer an inherently safer way of performing such reactions, partly due to the fact that minimal material needs to be under pressure at any one time. The design of the reactor can be such that high pressures can be maintained safely with the aid of a suitable pressure-regulating device.

Gaseous reactions form another class of reaction well suited to a flow regime, particularly if a hazardous gas (for example hydrogen, carbon monoxide or ozone) is generated *in situ*. Pressurising the flow stream allows gas to be readily dispersed and mixed with the liquid. This methodology has been well exploited in recent years as a safe and convenient way of performing hydrogenation reactions previously considered too dangerous to perform in a normal laboratory environment and provides a perfect example of many of the benefits of continuous flow chemistry.[12,13]

5.7 Packed Bed Reactors

The use of solid supports as a means of enhancing chemical synthesis has been advocated in recent years.[15] The convenience of handling reagents in this manner together with the advantages of offering cleaner chemistry, and its suitability for use in automated processes, has been exploited through the combinatorial chemistry era in the 1990's, and more recently in parallel or array-based chemistry. The initial concept of solid-phase chemistry, where the molecule of interest is built up whilst attached to a solid support, gave way to more broadly applicable techniques encompassing supported reagents and catalysts, and supported scavengers, where the target molecule remains in solution.

It has been recognised that each of these techniques is in itself ideal for operation in a continuous flow manner, with the solid support medium packed into a column and plumbed into the flow line at the required point in the process. In fact the whole concept of solid-supported chemistry was initially developed for peptide synthesis in the 1960's,[31] and many of the devices developed for this aim were operating in an automated multistep continuous flow manner.[32] Large surface areas created on the solid support allow high levels of contact between reactants, leading to rapid reaction profiles. Problems associated with scaling up solid-supported chemistry in a batch mode, mainly down to achieving efficient mixing without mechanically breaking the support material, and shortening its lifetime, can be avoided by operating in continuous flow.

The nature of the solid support plays a big role in its efficiency and effectiveness. Early supports tended to be porous polymers, for example polystyrene, requiring permeation of the solution phase into the support to gain access to all active sites. These materials also suffered from the fact that they swell or contract when in contact with organic solvents. This makes them difficult to contain within a sealed system, and causes problems with restricting flow rates. Silica particles have been used more recently as a heterogeneous support,[33] providing a more convenient medium for handling in a packed column. Active sites are more accessible and the material is not prone to state change in contact with organic solvents. Flow characteristics within the packed bed are an important consideration for optimum interaction of the stationary and mobile phases. Uniform, evenly distributed flow through the bed is required for reproducible results, which can be achieved with regular, evenly packed spherical particles. Minimising the particle size gives the maximum contact surface area and increases the loading capacity of the support, features that offer good efficiency. However, this can lead to the formation of a bed requiring high pumping pressures to push the mobile phase through, placing an additional load on the pumping system.

A new generation of solid supports has recently been developed, and found to be particularly well suited to continuous flow systems. Monolithic silica rods,[34–37] produced by a sol–gel process, form a series of continuous channels of precisely defined, rigid structure, offering large surface areas. The functionalisation of these channels with catalysts or reagents provides an activated flow channel with a low pressure drop giving improved mass transfer between catalyst and mobile phase. Porosity, composition and shape of the rod and channels can be readily varied during their manufacture, providing good flexibility for reactor design. Examples of monoliths functionalised with azide for use in continuous flow, including Curtius rearrangement reactions,[38] and Pd(0) for Heck reactions,[39] have been recently reported. Similarly, a range of chiral catalysts for asymmetric synthesis in continuous flow have been produced,[40] and higher selectivity obtained, compared to batch reactions and other solid supports. The importance of the nature of the support medium and variation in access to catalytic sites is demonstrated by a complete reversal of selectivity in a Diels–Alder cycloaddition on switching from a polymeric material to a monolithic column.

5.8   Published Chemistry Examples

Ever increasing numbers of publications are appearing in the scientific literature as interest in flow technology has led to increasing academic research and
uptake within industry over the last few years. Examples are presented in the
following section to illustrate some of the key topics. This is intended to give a
flavour for what might be possible, rather than an exhaustive review of the
subject; many more examples can be found for each category.

5.8.1   Hazardous Reagents

Reactions requiring the use of potentially hazardous reagents need extensive
development time to identify sufficiently safe operating conditions, or to design
alternative safer syntheses, particularly if they are to be performed on a significant scale. However, because the amounts of materials being handled at any
one time are relatively small and, with the additional features of enhanced
control of temperature and mixing, such reactions are often ideally suited to a
flow reactor. For example, the ring expansion reaction with ethyl diazoacetate
(Scheme 5.5) has been shown to be particularly problematic, and unlikely to be
performed in a batch reactor because of identified safety issues. These relate to
a delayed onset, highly exothermic reaction, vigorous evolution of large
quantities of gas and the presence of a thermally unstable diazo species.[41]

   However, performing this reaction in a flow reactor proved safe and
straightforward, with excellent control of the rapid reaction (1.8 min residence
time), giving an 89% yield of required product, with a throughput of 91 g h^{-1}.

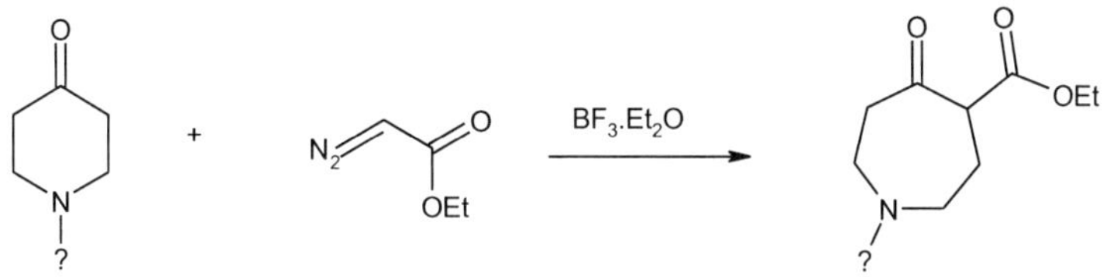

Scheme 5.5   Highly exothermic ring expansion reactions can be safely performed in a
flow reactor.

5.8.2   Exothermic Reactions

5.8.2.1   *Oxidation*

The TEMPO-catalysed oxidation of alcohols to aldehydes and ketones is an
important transformation in medicinal chemistry due to its selectivity between
primary and secondary alcohols. Typically, sodium hypochlorite is used as an
oxidant and, because of the exothermic nature of the reaction, its addition to an
organic solution of the alcohol has to be performed slowly over a period of
several hours. Highly efficient mechanical stirring is also required. Transfer of

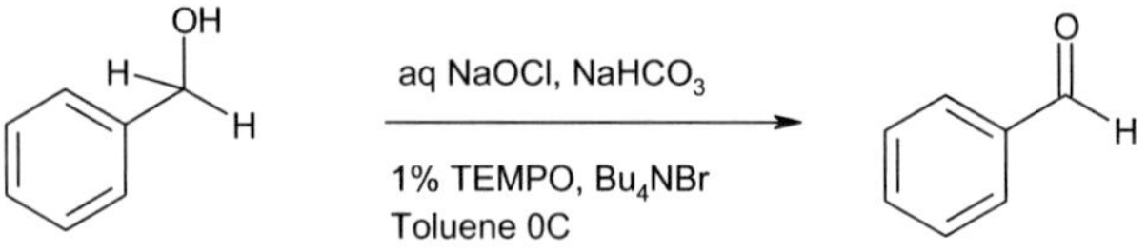

Scheme 5.6   TEMPO-catalysed oxidation.

this reaction to a continuous flow reactor has been demonstrated (Scheme 5.6),[42] and reactions requiring several hours in a batch reactor were reduced to 0.7–3 min residence time under flow conditions, whilst also increasing reaction yield.

5.8.2.2   *Fluorination*

The introduction of fluorine into a molecule is an important process for many pharmaceutical and agrochemical products but the ability to do this regioselectively is often a difficult challenge. Many fluorinating agents exist but the ability to use elemental fluorine directly to accomplish this would be highly desirable. However, this methodology has been severely restricted by the requirements for handling such hazardous materials and the inability to control such thermodynamically exothermic processes, particularly on a large scale. It has recently been shown that it is possible to perform this transformation in a continuous flow reactor.[43] The advantages of efficient heat transfer and the ability to minimise use of hazardous reagents has been demonstrated with direct fluorination of a series of β-keto esters with gaseous fluorine (Scheme 5.7).

Typical batch conditions for this process require continuous bubbling of a stream of fluorine gas over a long period of time through the bulk solution containing the β-keto ester. The continuous flow conditions developed involved pumping a stream of β-keto ester solution into one inlet channel, to meet and mix with a stream of fluorine gas at controlled flow rates. The high surface area created at the gas/liquid interface leads to the fast and efficient reaction observed. A higher selectivity for the required mono 2-substituted product over bis-fluorinated side products was observed with the flow reactor. Throughputs of several hundred grams have been demonstrated through continuous operation of the reactor, and extrapolation by scaling out to multiple channel systems predicts throughputs of several kg per day could readily be achieved.

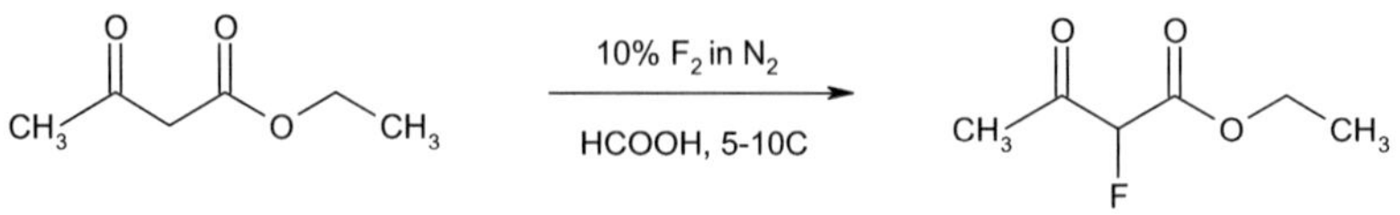

Scheme 5.7   Regioselective fluorination with fluorine gas has been performed in a flow reactor.

5.8.2.3 Carbamate formation

Formation of methyl carbamates from methyl chloroformate and an amine is generally an exothermic reaction; the batch preparation of *N*-methoxy-carbonyl-L-*tert*-leucine (Scheme 5.8) shows a significant temperature rise on addition of the methyl chloroformate.[41] Carefully controlled addition would be required to scale this reaction up.

Transfer of the conditions to a flow reactor enabled fast reaction times (7 min) to be achieved safely and in high yield through the efficient control of this exotherm, and a multi kg day^{-1} throughput was possible.

5.8.3 Unstable Intermediates

5.8.3.1 Diazonium Salts

The *in situ* generation of unstable diazonium salts followed by immediate quenching with a chlorinating agent within a flow reactor[44] nicely demonstrates the ability to safely handle unstable intermediates (Scheme 5.9). Here, the unstable species, normally requiring extensive precautions for its safe handling, is only present in small quantities at any one time within the fully contained flow assembly, and is consumed shortly after its generation.

5.8.3.2 Lithium Anions

The use of lithium anions, generated from butyl lithium (*n*-BuLi), is widespread in organic synthesis. However, the low stability of these species and the requirement to handle them at sub-zero temperatures makes the scale-up of such processes difficult to handle. The two step formation of 3-methoxy-phenyllithium, followed by its addition to cyclohexanone in a batch process,

Scheme 5.8 Methyl carbamate formation.

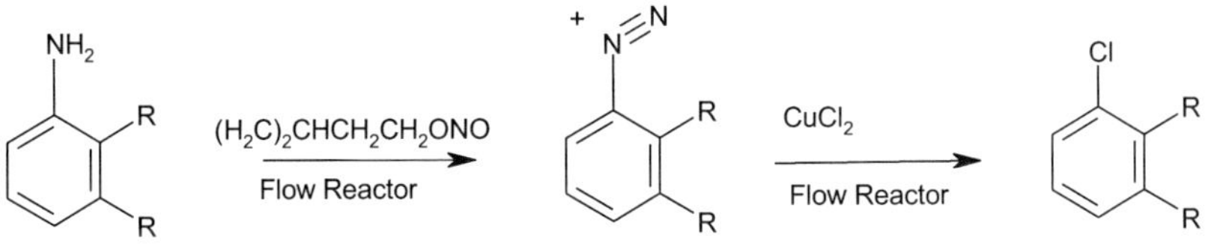

Scheme 5.9 Unstable intermediates can be immediately quenched in a flow reactor.

has been studied (Scheme 5.10).[41] Both steps are exothermic, giving high temperature rises, and a bulk temperature of –65 °C was required to give satisfactory yields (80%) of the required alcohol, due to the lack of stability of the aryllithium intermediate. Running this chemistry in a flow reactor clearly demonstrated advantages of superior temperature control and the ability to quickly form and subsequently react unstable intermediates, producing superior results. A residence time of merely 17 s was required to form the aryllithium species, at a reactor temperature of –14 °C; the resulting quench with cyclohexanone produced an improved yield (87%) for the alcohol.

5.8.4 Selectivity

The nitration of salicylic acid, using $HNO_3/AcOH$ (Scheme 5.11) has been shown to benefit from operating in a flow reactor.[45] Under specific conditions complete conversion of reactants could be achieved with a residence time of less than 7 min, with the formation of purely mononitro-derivatives, and high selectivity of the desired 5-nitrosalicylic acid.

Further examples of selective nitration under highly exothermic conditions prone to hazardous runaway situations, can be found.[46] Nitration of pyrazole 5-carboxylic acid, (Scheme 5.12) a key intermediate in the synthesis of

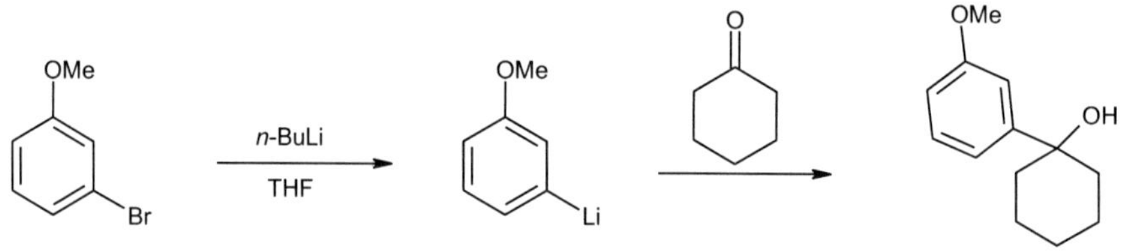

Scheme 5.10   Unstable aryllithium species give improved results in a flow reactor.

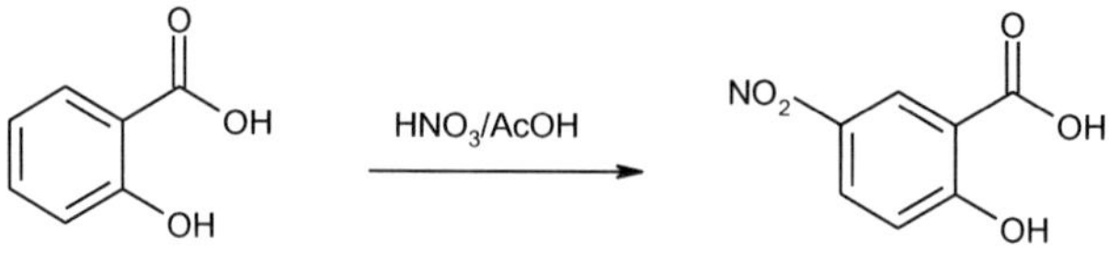

Scheme 5.11   Selective nitration can be safely performed in a flow reactor.

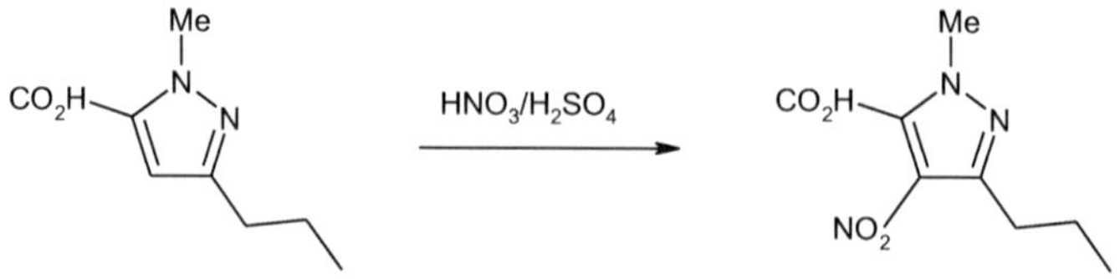

Scheme 5.12   Selective nitration benefits from precise temperature control.

Sildenafil®, has been shown to be temperature-sensitive for both synthesis and quench steps. Precise control of temperature is required for high chemoselectivity and to avoid product decomposition. The ideal reaction temperature was identified to be 90 °C, but exothermic decaboxylation was shown to occur at 100 °C, leading to loss of product. A dangerous reaction runaway and build-up of pressure from CO_2 produced were also observed.

To perform this reaction in batch mode required slow batchwise addition of nitrating agent followed by careful analysis, over a period of hours. Transferring this process to a continuous flow system not only allowed precise control of the temperature required to generate the product in optimum yield and purity but removed all hazardous operating conditions.

5.8.5 Multistep Processes

A recent example shows the convenience of linking multiple synthetic steps in a continuous process for formation of biologically active 1,2,4-oxadiazoles.[47] In this case a three-step batch process to create the target molecules is converted to a linked three-stage continuous process (Scheme 5.13).

The formation of amidoxime from aryl nitrile and hydroxylamine is achieved in the first reactor at 150 °C in 6 min. It was found necessary to cool the amidoxime to 0 °C as it exited the reactor, before it reached the point where acid chloride was introduced. Formation of O-acyl amidoxime occurs in the second reactor at room temperature in approximately 2 min, and the third reactor performs the dehydrative cyclisation to generate the required oxadiazole in 10 min at 200 °C and 9 bar pressure.

In this example, the overall yield for the process (45%) is equivalent in both batch and flow, but the major advantage is achieved through reducing process

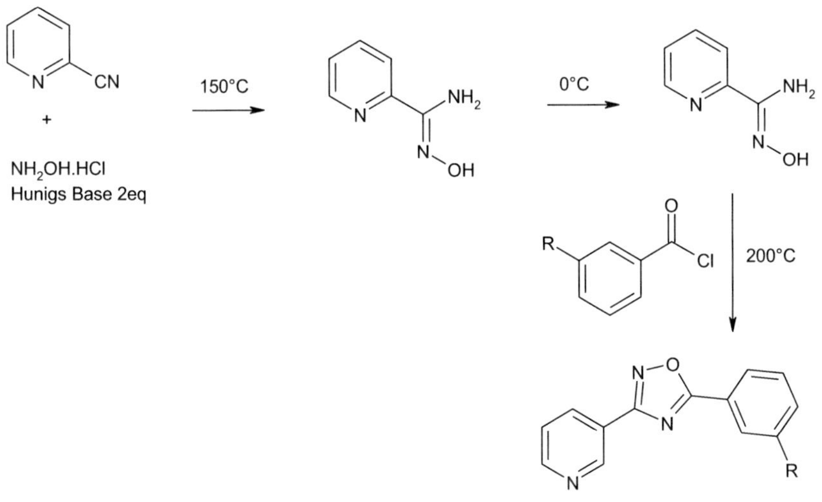

Scheme 5.13 Multistep synthesis of 1,2,4-oxadiazoles.

time from several days to around 30 min. Additionally, a potentially hazardous and unscalable high-temperature pressurised step in a sealed reactor has been replaced by a relatively safe flow reaction utilising superheated solvents at elevated pressures, a route that now becomes possible to scale up.

Other published examples of combined multistep sequences include an enolisation, oxidation and quench to form 6-hydroxybuspirone on a 47 kg scale,[48] a flexible route to either 4,5-disubstituted thiazoles or imidazoles using an immobilised catalyst,[49] and a three-step route to triazoles *via* oxidation, homologation and an azide–alkyne cycloaddition sequence.[50]

5.9 Integrating Continuous Flow with Other Technologies

5.9.1 Microwaves

The intriguing prospect of integrating two technologies that have both individually demonstrated substantial benefits, microwave enhanced synthesis and continuous synthesis, has already been touched on. Both have shown the ability to significantly impact on reaction rates, and subtly change the way chemistry is performed, opening up potential chemical space by developing "new chemistry". The question is: can the two techniques be harnessed to produce additive effects?

Two strategies have generally been employed to explore possibilities. Firstly, tube-based flow reactors have been built to fit within the cavity of existing commercial microwaves. Reactants flowing through the tubing can then be exposed to and energised by the microwave field in a similar way to that in the batch reactor. In this way microwave chemistry can be performed in a system not limited by scale, offering a means of overcoming what might previously have been considered a technology gap. Transfer of a synthesis to continuous flow microwave is not necessarily straight forward.[51] Generally, optimised batch microwave conditions are not suitable for flow, and need to be re-optimised, which can be a time-consuming and labour-intensive process. Existing microwave cavities have been tuned and focussed for batch reactors, and a non-uniform distribution of microwaves in the cavity can lead to hot spots and temperature gradients along the path of a flow reactor. The coupling of microwaves with the reaction solution will also be influenced in an unpredictable way by the movement of solution within the cavity. However, despite these technical issues, many examples have been published[52–58] of chemistry performed in this way, including nucleophilic aromatic substitution, Suzuki coupling, esterification, Wittig and ring-closing metathesis to name a few.

A second strategy involves treating the walls of the flow reactor with a thin layer of metal. In this case, microwave energy is directed at the walls of the reactor itself, rather than the solution flowing through it. Heat energy is subsequently transferred by normal conduction from the flow channel walls to the reaction mixture. Microwaves can be applied more consistently to the reaction by this means, but some of the advantages of direct coupling with the reaction mixture may be lost. Gold has been demonstrated as a suitable coating material for this application,[59,60] and the principle has been extended to incorporate a palladium coating, such that the reactor surface not only acts as a heat transfer mechanism, but also as a catalyst for the reaction.[61,62]

An alternative approach to combining flow technology with microwaves is the concept of automated stop flow.[63] Not truly a continuous flow process but, as the name suggests, a half-way house whereby reactants are pumped into the microwave cavity, where they are held in a batch reactor. Once the reactor is full, input flow is stopped and microwave energy applied to the reactor for the required length of time just as it would be in a standard batch microwave. On completion of reaction, product mixture is pumped from the reactor to a collection vessel, before refilling with a fresh charge of reactants. Many of the problems associated with attempting to apply microwaves directly to a continuous flow stream can be avoided in this way.

5.9.2 Other Technologies

Similar efforts are ongoing to employ other technologies currently used in batch processing in a continuous flow manner. Reaction activation by ultrasound[64–66] and photolysis[67] are two such examples, although these have less widespread application than microwaves, so have received less attention. A further example utilises "ohmic heating", whereby precise heating control can be achieved along the length of a flow reactor with the application of an electrical current to the channel walls from an accurately controlled power supply.[68] The ability to collect thermal feedback from this device gives the additional benefit of obtaining analytical insight into the course of the reaction.

5.10 Beyond Synthesis

Although the major focus on developing continuous flow technologies has naturally been directed at synthetic chemistry applications, it has already been noted that multistep processing, either linking sequential synthetic steps or linking synthesis with post-synthesis operations such as work-up and purification, is highly desirable. To this aim, it has been demonstrated that several post-synthetic processes lend themselves well to a continuous flow regime.

5.10.1 Liquid–Liquid Extraction

Handling biphasic components in a continuous flow set-up has been well established in synthetic applications, where liquid–gas and liquid–liquid systems have been developed.[69] The standard methodology for achieving this is to introduce the two phases at a simple fluidic mixer device where they can combine in a variety of ways depending on a range of factors. The two common biphasic flow patterns observed are either parallel (or laminar) flow where the two phases lie side by side, or segmented flow where they form discrete alternating bubbles. The formation of bubbles can be easily achieved; their size and shape depend on such factors as the diameter of the flow channel, relative flow rates and physical properties of both phases, and the reactor material. The attractive feature of these bubbles is that they produce a very high surface area to volume ratio, and therefore maximise contact between phases. In a synthetic application this can result in dramatically increased reaction rates, but high contact area is also an ideal scenario for achieving rapid transfer of solutes from one phase to the other, lending itself well to the requirements of liquid–liquid extraction. Ideally the size of bubbles should be minimised for the optimum effect. An additional feature, created by the interfacial tension and surface energies present in the flowing system, is the formation of internal vortices within the individual bubbles. This creates a perfect environment for rapid mass transfer between phases, which has been exploited in both reaction work-up and log D solubility studies.[70]

5.10.2 Purification

By far the most common method of purification employed by the chemist, particularly on a research scale, is chromatography and, to exploit the full range of standard techniques, the ability to integrate generic chromatography into a flow process would be considered essential. However, in spite of the fact that standard column chromatography is in itself a flow process, this is one of the areas that is currently least well addressed. The difficulty lies in maintaining differential adsorption rates of components in the mixture with the solid support whilst continually introducing new material to the top of the column. Strategies employing supported scavengers have been successfully demonstrated,[15] but these utilise specific interactions for targeting known impurities, and do not lend themselves to generic methods.

One possible approach to the problem requires a series of chromatography columns with an automated switching valve linking them, such that one column can be loaded with the flow stream direct from the reactor before switching to a second column. While the second column is being loaded the first column is processed in the standard way. With careful timing of column switching and elution, followed by column washing, it can be envisaged that a continuous purification method could be handled in this way. One further complicating factor to note is that solvent employed for the reaction must also be compatible with the chromatographic method. A highly polar reaction solvent like dimethylformamide is likely to be detrimental to any purification process, so planning ahead would be required for this approach to be successful.

An extension of this concept is continuous simulated moving bed chromatography (CSMBC) which employs an arrangement of columns and switching valves and allows a continuous separation process to operate. This process has been successfully used in the separation of binary mixtures, for example racemates,[71] but is not suited to generic separations of multicomponent mixtures.

One technique that has been shown to offer potential for genuine generic continuous chromatography is continuous annular chromatography (CAC).[72–75] This features a revolving annular packed bed onto which the mixture to be separated is continuously loaded and eluted. Careful timing is required such that total elution time of the reaction mixture is equal to or less than one complete revolution of the annular column. Although the method shows promise, it has yet to be fully established in an industrial setting.

5.10.3 Continuous Flow Crystallisation

It was noted previously that one of the biggest practical hurdles to overcome in developing a continuous flow chemical synthesis was the reliance on the solubility of the reagents and products. However, an interesting deviation from this is that continuous flow crystallisation has been shown to be an extremely good way of controlling particle size and particle size distribution. A small particle size with a narrow distribution range can be an important feature for efficient *in vivo* delivery of a pharmaceutical product, and hence there is much interest in being able to achieve this, particularly in a manufacturing environment.

The control of crystallisation induced by co-solvent addition or careful control of temperature gradient profiles is significantly enhanced when the general process parameters can be carefully controlled. Further control of nucleation and crystal growth, for example by the application of ultrasound has also shown to be particularly beneficial.

Examples of crystallisation applied to continuous processes have been described in the literature for a pharmaceutically active steroid developed as a candidate for asthma and irritable bowel syndrome.[76]

More recently the use of continuous flow principles for nanoparticle formation, where regular spherical particles are required, has been shown to be advantageous.[77] Reliable and reproducible generation of the desired particle formulation, with a tighter size distribution than can be achieved in batch is possible using a microfluidic approach.

5.11 Conclusion

One of the most exciting applications for continuous flow is the possibility for "closed loop" drug discovery. The concept of performing an internally controlled "intelligent" cycle whereby an operation is performed, results analysed, and the system responds to determine the next operation may be the future of the drug discovery process. Such cyclical processes could be envisaged being applied to reaction optimisation and process improvement where sequential reactions are performed to optimise factors such as yield, purity and side-product formation. Ultimately drug discovery itself could be enabled by combining synthesis and biological screening on one continuous cyclical platform.[78] The realisation of this concept may be some way off, but the principal idea of "lab on a chip" where all operations of chemical synthesis and biological screening can be combined on one small microchip has been around for some while.[79]

Automated "closed loop" reaction optimisation has already been described in the literature.[80] Integration of an analytical HPLC system into the output flow from a continuous reactor allows a sample to be removed and analysed to determine its composition. The ability to easily vary reaction parameters such as temperature, flow rate (residence time) and reagent stoichiometry in an automated way gives the opportunity to assess the effect of these changes on product composition. The use of software to interpret results and predict the next parameter change to improve conditions closes the optimisation loop.

Continuous flow biological screening processes are well developed,[81–83] so it might appear a simple step to combine the output of a chemical reactor with the biological ingredients of a flow screening process. However, issues such as the massive differences in concentration and scale, and the incompatibility of solvents between a chemical and biological process, leads to significant problems. Developing a software package intelligent enough to interpret biological screening results relative to chemical structure, with the ability to define the next compound to make in the sequence, is also a tall order but maybe as all the relevant technologies evolve one day the ultimate automated closed loop drug discovery system may be achieved.

References

1. V. Hessel, P. Lob and H. Lowe, *Curr. Org. Chem.*, 2005, **9**, 765–787.
2. V. Hessel and H. Lowe, *Chem. Eng. Technol.*, 2005, **28**, 267–284.
3. A. M. Thayer, *Chem. Eng. News*, 2005, **83**, 43–52.
4. J. R. Bourne, J. Lenzner and S. Petrozzi, *Ind. Eng. Chem. Res.*, 1992, **31**, 1216–1222.
5. R. A. Taylor, W. R. Penney and H. X. Vo, *Ind. Eng. Chem. Res.*, 2005, **44**, 6095–6102.
6. M. Patel and G. Gasparini, *Spec. Chem.*, 2009, **4**, 22–23.
7. G. Mueller, *Chem. Files*, **5**(7), 15–16.
8. J. Yoshida, A. Nagaki, T. Iwasaki and S. Suga, *Chem. Eng. Technol.*, 2005, **28**, 259–266.
9. W. Pringle, *Tetrahedron Lett.*, 2008, **49**, 5047–5049.
10. C. O. Kappe, *Angew. Chem., Int. Ed.*, 2004, **43**, 6250–6284.
11. T. N. Glasnov, S. Findenig and C. O. Kappe, *Chem.–Eur. J.*, 2009, **15**, 1001–1010.
12. R. V. Jones, L. Godorhazy, N. Varga, D. Szalay, L. Urge and F. Darvas, *J. Comb. Chem.*, 2006, **8**, 110–116.
13. T. Razzaq, T. N. Glasnov and C. O. Kappe, *Eur. J. Org. Chem.*, 2009, **9**, 1321–1325.
14. V. Franckevicius, K. R. Knudsen, M. Ladlow, D. A. Longbottom and S. V. Ley, *Synlett*, 2006, **6**, 889–892.

15. I. R. Baxendale, R. I. Storer and S. V. Ley, in *Polymeric Materials in Organic Synthesis and Catalysis*, ed. M. R. Buchmeiser, Wiley-VCH, Weinheim, 2003, pp. 53–136.
16. I. R. Baxendale, J. Deeley, C. M. Griffiths-Jones, S. V. Ley, S. Saaby and G. K. Tranmer, *Chem. Commun.*, 2006, 2566–2568.
17. I. R. Baxendale, C. M. Griffiths-Jones, S. V. Ley and G. K. Tranmer, *Synlett*, 2006, **3**, 427–430.
18. T. Schwabe, V. Autze, M. Hohmann and W. Stirner, *Org. Process Res. Dev.*, 2004, **8**, 440–454.
19. C. O'Driscoll, *Chem. Ind.*, 26 March 2007, 5–6.
20. R. Ashe, *Manufacturing Chemist,* Sept., 2008, 83–84.
21. B. Wilson, D. C. Sherrington and X. Ni, *Ind. Eng. Chem. Res.*, 2005, **44**, 8663–8670.
22. H. Wakami and J. Yoshida, *Org. Process Res. Dev.*, 2005, **9**, 787–791.
23. R. C. Wheeler, O. Benali, M. J. Deal, E. Farrant, S. J. F. MacDonald and B. H. Warrington, *Org. Process Res. Dev.*, 2007, **11**, 704–710.
24. K. Golbig, A. Kursawe, M. Hohmann, S. Taghavi-Moghadam and T. Schwalbe, *Chem. Eng. Commun.*, 2005, **192**, 620–629.
25. K. Golbig, M. Hohmann, A. Kursawe and T. Schwalbe, *Chem. Eng. Commun.*, 2004, **76**, 598–603.
26. V. Skelton, G. M. Greenway, S. J. Haswell, P. Styring, D. O. Morgan, B. H. Warrington and S. Y. F. Wong, *Analyst*, 2001, **126**, 11–13.
27. C. Wiles, P. Watts and S. Haswell, *Org. Process Res. Dev.*, 2004, **8**, 28–32.
28. N. Nikbin and P. Watts, *Org. Process Res. Dev.*, 2004, **8**, 942–944.
29. C. Wiles, P. Watts and S. J. Haswell, *Chem. Commun.*, 2007, 966–968.
30. D. M. Roberge, L. Ducry, N. Bieler, P. Cretton and B. Zimmermann, *Chem. Eng. Technol.*, 2005, **28**, 318–323.
31. R. B. Merrifield, *J. Am. Chem. Soc.*, 1963, **85**, 2149–2154.
32. V. Krchnak, J. Vagner and O. Mach, *Int. J. Pept. Protein Res.*, 1989, **33**, 219–213.
33. A. J. Sandee, D. G. I. Petra, J. N. H. Reek, P. C. J. Kamer and P. W. N. M. Leeuwen, *Chem.–Eur. J.*, 2001, **7**, 1202–1208.
34. M. Merhar, A. Podgornik, M. Barut, A. Strancar and M. Zigon, *J. Sep. Sci.*, 2003, **26**, 322–330.
35. G. Jas and A. Kirschning, *Chem.–Eur. J.*, 2003, **9**, 5708–5723.
36. A. Kirschning, W. Solodenko and K. Mennecke, *Chem.–Eur. J.*, 2006, **12**, 5972–5990.
37. U. Kunz, A. Kirschning, H. Wen, W. Solodenko, R Cecilia, C. O. Kappe and T. Turek, *Catal. Today*, 2005, **105**, 318–324.
38. M. Bauman, I. R. Baxendale, S. V. Ley, N. Nikbin and C. D. Smith, *Org. Biomol. Chem.*, 2008, **6**, 1587–1593.
39. N. Nikbin, M. Ladlow and S. V. Ley, *Org. Process Res. Dev.*, 2007, **11**, 458–462.
40. X. Y. Mak, P. Laurino and P. H. Seeberger, *Beilstein J. Org. Chem.*, 2009, **5**, 19.

41. X. Zhang, S. Stefanick and F. J. Villani, *Org. Process Res. Dev.*, 2004, **8**, 455–460.
42. P. D. Hampton, M. D. Whealan, L. M. Roberts, A. A. Yaeger and R. Boydson, *Org. Process Res. Dev.*, 2008, **12**, 946–949.
43. R. D. Chambers, M. A. Fox, D. Holling, T. Nakado, T. Okazoe and G. Sandford, *Lab Chip*, 2005, **5**, 191–198.
44. R. Fortt, R. C. R. Wooton and A. J. de Mello, *Org. Process Res. Dev.*, 2003, **7**, 762–768.
45. A. A. Kulkarni, N. T. Nivangune, V. S. Kalyani, R. A. Joshi and R. R. Joshi, *Org. Process Res. Dev.*, 2008, **12**, 995–1000.
46. G. Panke, T. Schwalbe, W. Stirer, S. Taghavi-Moghadam and G. Wille, *Synthesis*, 2003, **18**, 2827–2830.
47. D. Grant, R. Dahl and N. D. P. Cosford, *J. Org. Chem.*, 2008, **73**, 7219–7223.
48. T. L. LaPorte, H. Hamedi, J. S. DePue, L. Shen, D. Watson and D. Hsieh, *Org. Process Res. Dev.*, 2008, **12**, 956–966.
49. I. R. Baxendale, S. V. Ley, C. D. Smith, L. Tamborini and A.-F. Voica, *J. Comb. Chem.*, 2008, **10**, 851–857.
50. I. R. Baxendale, S. V. Ley, A. C. Mansfield and C. D. Smith, *Angew. Chem., Int. Ed.*, 2009, **48**, 4017–4021.
51. O. Benali, M. J. Deal, E. Farrant, D. Tapolczay and R. Wheeler, *Org. Process Res. Dev.*, 2008, **12**, 1007–1011.
52. E. Comer and M. G. Organ, *J. Am. Chem. Soc.*, 2005, **127**, 8160–8167.
53. N. S. Wilson, C. R. Sarko and G. P. Roth, *Org. Process Res. Dev.*, 2004, **8**, 535–538.
54. T. Schwalbe and K. Simons, *Chem. Today*, 2006, **24**, 56–61.
55. W. S. Bremner and M. G. Organ, *J. Comb. Chem.*, 2007, **9**, 14–16.
56. T. Calewiski, A. Faux and C. Strauss, *J. Org. Chem.*, 1994, **59**, 3408–3412.
57. P. He, S. Haswell and P. D. I. Fletcher, *Lab Chip*, 2004, **4**, 38–41.
58. M. C. Bagley, R. L. Jenkins, M. Caterina Lubinu, C. Mason and R. Wood, *J. Org. Chem.*, 2005, **70**, 7003–7006.
59. I. R. Baxendale, J. J. Hayward and S. V. Ley, *Comb. Chem. High Throughput Screen.*, 2007, **10**, 802–836.
60. P. He, S. Haswell and P. D. I. Fletcher, *Appl. Catal.*, 2004, **274**, 111–114.
61. G. Shore, S. Morin and M. G. Organ, *Angew. Chem., Int. Ed.*, 2006, **45**, 2761–2780.
62. B. K. Singh, N. Kaval, S. Tomar, E. Van der Eycken and V. S. Parmar, *Org. Process Res. Dev.*, 2008, **12**, 468–474.
63. J. D. Mosely and E. K. Woodman, *Org. Process Res. Dev.*, 2008, **12**, 967–981.
64. G. Cravotto, S. DiCarlo, M. Curini and V. Tumiati, *J. Chem. Technol. Biotechnol.*, 2007, **82**, 205–208.
65. J. Gao, K. D. Hungenberg and A. Penlidis, in *Modern Styrenic Polymers*, ed. J. Scheirs and D. Priddy, Wiley, Chichester, England, 2003, p. 106.
66. J. Gonzales-Garcia, C. Banks, B. Sljukic and R. Compton, *Ultrason. Sonochem.*, 2007, **14**, 405–412.

67. B. D. A. Hook, W. Dohle, P. R. Hirst, M. Pickworth, M. B. Berry and K. I. Booker-Milburn, *J. Org. Chem.*, 2005, **70**, 7558–7564.
68. C. A. Nielson, R. W. Chrisman, R. E. LaPointe and T. E. Miller Jr., *Anal. Chem.*, 2002, **74**, 3112–3117.
69. B. Ahmed, D. Barrow and T. Wirth, *Adv. Synth. Catal.*, 2006, **348**, 1043–1048.
70. M. Atimuddin, D. Grant, D. Bulloch, N. Lee, M. Peacock and R. Dahl, *J. Med. Chem.*, 2008, **51**, 5140–5142.
71. J. Strube, S. Haumreisser, H. Schmidt-Traube, M. Schulte and R. Ditz, *Org. Process Res. Dev.*, 1998, **2**, 305–319.
72. J. Wolfgang and A. Prior, in *Modern Advances in Chromatography*, ed. R. Freitag, Springer, Berlin, Heidelberg, New York, 2002, vol. 76, pp. 233–255.
73. H.-J. Bart, J. Brozio and R. Herbsthofer, *J. Therm. Sci.*, 2000, **9**, 129–134.
74. B. Finke, B. Stahl, M. Pritschet, D. Facius, J. Wolfgang and G. Boehm, *J. Agric. Food Chem.*, 2002, **50**, 4743–4748.
75. A. Thiele, T. Falk, L. Tobiska and A. Seidel-Morgenstern, *Comput. Chem. Eng.*, 2001, **25**, 1089–1101.
76. J. G. Van Alsten, L. M. Reeder, C. L. Stanchina and D. J. Knoecgel, *Org. Process Res. Dev.*, 2008, **12**, 989–994.
77. A. Jahn, J. E. Reiner, W. N. Vreeland, D. L. DeVoe, L. E. Locascio and M. Gaitan, *J. Nanoparticle Res.*, 2008, **10**, 925–934.
78. S. Y. F. Wong-Hawkes, J. C. Matteo, B. H. Warrington and J. D. White, in *New Avenues to Efficient Chemical Synthesis, Emerging Technologies*, ed. P. Seeberger, T. Blume, Springer Berlin, Heidelberg, 2006, pp. 39–55 .
79. B. H. Weigl, R. L. Bardell and C. R. Cabrera, *Adv. Drug Delivery Rev.*, 2003, **55**, 349–377.
80. M. Hawes, *Speciality Chemicals Magazine*, 2006, pp. 52–53.
81. A. M. Clark, K. M. Sousa, C. Jennings, O. A. MacDougald and R. T. Kennedy, *Anal. Chem.*, 2009, **81**, 2350–2356.
82. A. M. J. Law and M. D. Aitken, *Appl. Environ. Microbiol.*, 2005, **71**, 3137–3143.
83. A. R. de Boer, B. Bruyneel, J. G. Krabbe, H. Lingeman, W. M. A. Niessen and H. Irth, *Lab Chip*, 2005, **5**, 1286–1292.

CHAPTER 6
Emerging Synthetic Technologies

BB Consultants Ltd, 45 The Drive, Hertford, SG14 3DE, UK

6.1 Introduction

The generation of novel products in the pharmaceutical industry has always
taken many years. A time from concept or screen to market of 12 or so years is
not unusual during which time the ideas and technologies underpinning drug
discovery can change considerably. As a result it has usually been difficult to
correlate discovery methodology with commercial success. Only now are we in
a position to see with some clarity that the move which started in the early
1990's from rational iterative design towards high throughput serendipity
screening (HTS) of large numbers of compounds has had limited success. The
screening of historic molecule collections further amplified by the proactive
high throughput synthesis of new congeners based on combinatorial principles,
has not delivered the anticipated increase in new targets, novel drugs, and
enhanced profitability, despite a 3-fold increase in the number of molecules
entering development. For example, at the end of 2006 the US Food and Drug
Administration (FDA) pointed to a near doubling of US research and development (R&D) costs between 1993 and 2004, whilst the number of approval
applications had shown only a 38% increase with only 7% for novel compounds.[1,2] Further evidence that reduced output may be associated with
reduced novelty are reports showing that market exclusivity has decreased from
5 years to about 2 months and that during the period 2000–2008 52% of dis-
continued molecules were dropped for 'financial' and or 'strategic' reasons,[3,4]
usually implying that marketability has been impacted by unexpected

RSC Drug Discovery Series No. 11
New Synthetic Technologies in Medicinal Chemistry
Edited by Elizabeth Farrant
© Royal Society of Chemistry 2012
Published by the Royal Society of Chemistry, www.rsc.org

competitors with similar products. The general move back to more target-focussed teams and proven strategies as a remedy is typified by the recent remarks of GlaxoSmithKline's (GSK's) CEO.[5] Accordingly this chapter focuses on emerging tools that can assist this process.

It is understandable how the "industrial screening (HTS)" process of screening stored single compounds could both curtail the discovery of novel leads yet assist a development candidate's ability to pass the hurdles presented by the pre-clinical phase. Currently, cost, availability and equipment limitations in the procurement, storage, dispensing and curation limit HTS collections to a few million structures, which is infinitesimal relative to the 10^{63} distinct chemotypes possible in drug-like chemical space,[6,7] each of which will have many variants. However, this collection must serve as the source of all leads for all the proteins that will be screened. Inevitably the collection can only contain "foreseen" compounds and the selection of compounds to be included will be biased by historic notions of "permissive" drug structures noted for high bioavailability and low toxicity, often coupled with synthetic ease. It is therefore likely that the triage processes of hit selection, lead selection, optimisation and candidate selection will converge on a compound historical chemotype well suited to survive pre-clinical triage. The likelihood of a new class approval, chemical "breakthrough" or "blockbuster" or an unopposed market is low. Accordingly, this chapter will focus on the iterative strategies that underwrote the period of highest innovation and fastest growth in the pharmaceutical industry, but avoid the cost and seeding issues of this historic paradigm.

6.2 Template Guided Systems

6.2.1 Overview

A fixed screening collection cannot contain every structural nuance within just one chemotype, never mind the myriad possible drug-like chemotypes. Whilst the activity of most drugs can be vitiated or severely reduced by just a very minor chemical change, the chances of finding the "best possible" lead in all of chemical space, or even a reliable pointer to it, is extremely small. The chances of success could be improved if the biological target could somehow be used to act as a template to guide the selective synthesis of its own ligand. Notionally, relative to synthesising a speculative library, this would also reduce the number of compounds needing to be synthesised. The organisational downside would be the absence of the extraneous compounds in the corporate compound collection, but arguably this would be better stocked by a planned diversity campaign.

This section examines three template-guided systems. Dynamic combinatorial libraries and "click" chemistry both seek to use the natural features of the target to guide the production of suitable ligands. In a sense they may be seen as internally iterative systems because of their trial and error approach. The third technology, based on siRNA displaying a binding motif for an intracellular target, attempts to confer systemic delivery and target specificity to these small RNAs.

6.2.2 Dynamic Combinatorial Chemistry

An early attempt to systemise the discovery of therapeutic small organic molecules was through the use of the combinatorial ("mix and split") approaches similar to those that had proved successful in genomics.[8-11] Because some reagent combinations would be more favoured than others, leading to uneven concentrations of components, library synthesis could rarely be conducted in solution as identification of "hits" would be extremely difficult, especially if partially reacted reagents survived. To avoid chaos, the components of combinatorial libraries were nearly always constructed as bead-bound substrates and, as no "purification" step (of a mixture) was possible, large excesses of reagents were used to ensure near quantitative yields. A notional example of the synthesis of a mixture of a large number of molecules derived from a much more limited set of units connected by successive and repetitive application of specific chemical reactions is shown in Figure 6.1.

Although the illustrated procedure appears facile, it must be remembered that prior to library synthesis extensive "validation" will have taken place of all the reactions involved to ensure that persistently low-yielding reagent combinations were eliminated. In addition a tedious quality-assurance program will have been conducted to ensure the quality of the library at all stages. It can take a year to plan, validate, prepare and check a large library. In practice the specialised cleavage and deconvolution techniques needed to screen bead-bound libraries proved to be frail processes with much resource being wasted on the pursuit of 'phantom' leads. Applying tags, often a short peptide sequence or a readily identified mass spectrum auxiliary, to aid the screening hit identification process simply added to the burden of library construction and even further restriction of usable chemistries.

Medicinal chemists have always attempted to save time by reacting to biological data with the design of a small number of separate compounds prepared in parallel specifically designed to enhance the ongoing elucidation of

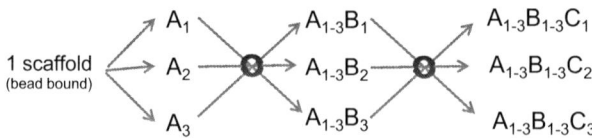

Figure 6.1 Combinatorial principles. A bead-bound scaffold is split 3 ways and each pool is treated with a different reagent to yield 3 new bead-bound compounds "A_n". These beads are then mixed and then split 3 ways and each new pool is treated with a different reagent to yield 9 new bead-bound compounds "$A_{1-3}B_{1-3}$". This process is repeated using 3 further reagents to give 27 bead-bound compounds "$A_{1-3}B_{1-3}C_{1-3}$".

structure–activity relationships (SAR) relating to the specific target. The numbers that could be managed were limited by human ability to deal with what might be several different synthetic routes. It was the advent of high throughput combinatorial and screening methodologies that founded the notion that a chemist's average productivity could be raised from less than a hundred compounds *per annum* to hundreds of thousands by producing many compounds from the same basic set of reactions. In this way it was anticipated that drug-like chemical space could be speculatively explored using large libraries of compounds as a common resource for detecting leads for "all" biological targets.

Although this methodology had been quite successful in biology when seeking, say, a particular pentapeptide in a combinatorial mixture (*i.e.* a large but finite linear set), it did not produce the hoped-for results when applied to the infinite continuum of 3-dimensional chemical space. Even within the space that could be covered there was a conflict. The easiest way to achieve large numbers of compounds was to use facile reactions to decorate many positions on a large scaffold with range of common small substituents, but the large common structure component compromised diversity. However, targeting significant numbers of compounds with very minimal common structure (*e.g.* a direct link or amide link, *etc.*) forces a move to produce a large number of elaborate "substituents" and eliminates the effectiveness of the approach.

An alternative approach is to direct a combinatorial approach towards a particular biological target. Dynamic combinatorial chemistry[12,13] is a method for conducting a target-specific lead search. The method relies on a continuous connection/de-connection process which can deliver a library containing all possible combinations of a set of reagents, but with little or no free reagent being present. The addition of a host molecule (*e.g.* receptor) or a guest molecule (*e.g.* a substrate) which can selectively bind a particular library member as a ligand will remove it from the equilibrium and mass action will result in it being replenished. The amount of captured ligand will therefore be amplified until either the relevant building blocks are exhausted, or the host or guest is saturated, or a state of equilibrium is reached with other building blocks. The supramolecular entity (assuming there is one predominant ligand) can then be removed and its components identified. Thus a dynamic combinatorial search can be conducted in two ways, "casting" and "moulding" (Figure 6.2), depending on whether a receptor or a substrate acts as the template which induces the assembly of a particular set of library components. The former is closest to fulfilling a lead discovery role. The latter is more appropriate for the discovery of catalysts and artificial receptors which have a less direct role in lead discovery.

Various chemistries have been examined as the basis of combinatorial libraries (Figure 6.3).[14,15] The earliest methods involved exchanges about a C=N bond involving imines,[16] hydrazones[17] and oximes.[18] Other methods have included disulphide exchange,[15] transesterification,[19,20] peptide bond exchange[21] and olefin metathesis.[22–24] Examples of sequential and simultaneous use of orthogonal library chemistries have also been reported.[25–27]

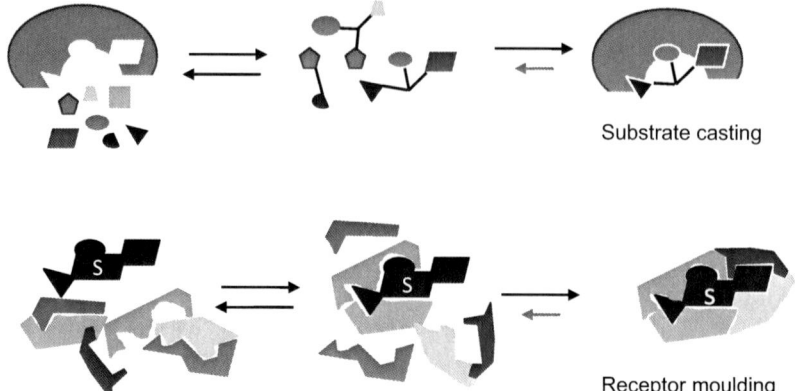

Figure 6.2 (upper) Schematic representation of "casting" (receptor induced selection) and (lower) "moulding" (selective self-assembly of the complementary "receptor" around a substrate (S)).

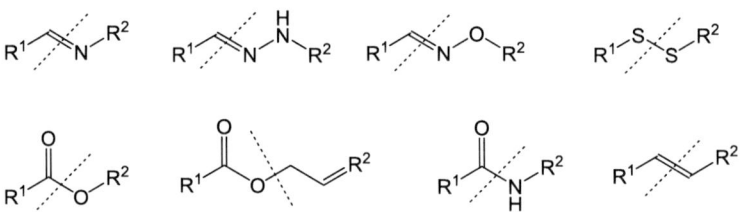

Figure 6.3 Interchangeable moieties of dynamic combinatorial libraries. Dotted line shows cleavage point for imines,[16] hydrazones, oximes, disulphides, esters, allyl esters, amides (peptides) and in olefin metathesis.

The methodological limitations of the approach are that continuous remodelling can occur without being skewed by a few highly preferred pairings or by the precipitation of many components. The benign conditions are needed to ensure that the integrity of the protein target is at odds with the reactivity required of the combining fragments. In addition, at the termination of the experiment the captured ligand is essentially "unstable", which may limit its value as a drug lead, and usually it must be stabilised by a further reaction to give a stable entity (which may not be a lead) for isolation and identification. Although selection and possibly amplification from a dynamic combinatorial library has been demonstrated,[15,28] no drug has yet ensued. In practice, the target-guided stepwise approach[29] in which the combinatorial fragments are first selected by their binding potential then joined by flexible linkers has possibly been more successful.

Moulding suffers fewer of these restrictions, since the template (perhaps an enzyme substrate) can be a more robust chemical entity than a protein, and most of the more recent work in this area has concentrated on this aspect.[14]

The main value of dynamic combinatorial libraries in drug discovery may therefore be to define specific catalysts as auxiliaries in other strategies or by providing a mould with greater electronic resolution and surface character-isation than a simple molecular imprint as a tool to capture bioisosteres of the original template.

6.2.3 Click Chemistry

Click chemistry signifies a paradigm in which molecules are assembled from smaller modular units under facile and benign conditions and with high atom economy.[30,31] Multistep or combinatorial approaches can access a great variety of structural geometries in which "click" substructures may be repeated many times. The proponents of the technology perceive this as a replication of the behaviour on amino acids in the formation of peptides. Although there are several reaction types (Figure 6.4) which have the thermodynamic driving force to deliver a stable product near quantitatively under mild conditions (*e.g.* urea formation, nucleophilic substitution into strained systems such as epoxides and

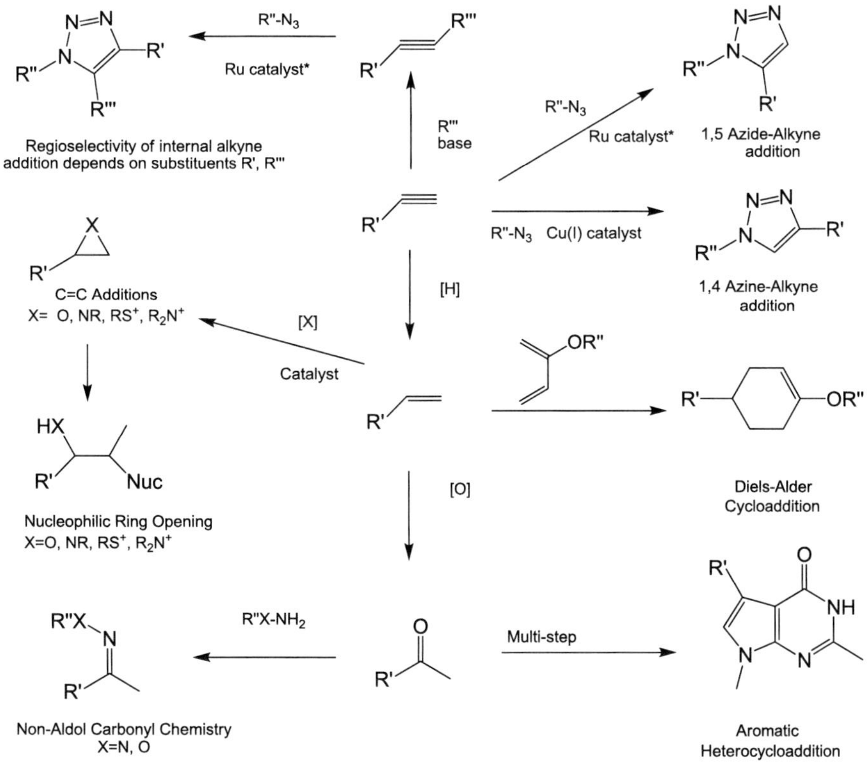

Figure 6.4   An extended scheme of "click" reaction types. * Ru catalyst is chloro-(pentamethylcyclopentadienyl) bis(triphenylphosphine)ruthenium(II).

aziridines, additions across a double bond or Diels–Alder cycloadditions), the literature of "click" is dominated by improved versions of the Huisgen 1,3-dipolar azide–alkyne cycloaddition.[32] For example a Cu(I) catalyst is used to achieve 1,2,3-triazoles with 1,4-disubstitution[33] or a Ru(II) catalyst to yield the 1,5 isomer,[34] each with regiospecific purity. Unlike the Cu(I)-catalysed reaction, which is limited to terminal alkynes, the Ru(II)-catalysed reaction can also process internal alkynes to yield fully substituted 1,2,3-triazoles.[34]

The undemanding reaction conditions for the azide–alkyne cycloaddition with its high yields of easily isolated products render it particularly adaptable to high throughput methodologies, including "lab-on-a-chip" formats,[35] and the products have been popular candidates for the speculative synthesis of corporate screening collections. This use has been facilitated by new preparations of organic azides in which unactivated olefins yield alkyl azides in the presence of a cobalt catalyst prepared *in situ* from a Schiff base ligand and $Co(BF_4)_2 \cdot 6H_2O$. Optionally, this reaction can be coupled to the Sharpless cycloaddition to yield the 1,4-triazole in a one-pot process.[36] Unfortunately, there appears to be no report showing that "click" collection components furnish validated leads or drugs beyond their relative abundance in the collection. A more powerful application has been as integral components within a discovery activity involving the biological target.

A variety of methods[8,37–39] has been used to explore a form of target-guided synthesis where chemistry is carried out in the presence of the biological target, which may then act as a template for the formation of its own ligand. "Click" chemistry, as exemplified by the formation of a triazole by azide–alkyne cycloaddition, is one of the few chemistries that pose only a low level of threat to the integrity of the protein target because suitably catalysed reactions can occur under near physiological conditions. The biological target acts to select a suitable azide–alkyne combination and holds them in close proximity to catalyse the irreversible formation of a triazole.[37,40] The general protocol[41] is to assemble sets of diverse azides and alkynes then dispense them pair-wise to form a complete combinatorial matrix in multiwell plates. The biological template, usually an enzyme, is added to every well and the wells are screened to detect the inhibitory activity. Wells showing activity are validated by comparison with another well containing the same azide–alkyne mixture but no enzyme using HPLC–mass spectrometry analysis. This method has been used for the *in situ* assembly of acetylcholinesterase (AChE) inhibitors with up to femtomolar potency[42–44] and crystal structure studies of selected ligand–AChE complexes demonstrated that the newly formed triazole was a strong contributor to binding.[42] However, this was not always the case as the same workers observed no significant contribution from the triazole group when carbonic anhydrase II was the target.[44]

The method seems to be a general strategy extendable to most enzymes possessing extended active sites or multiple binding pockets and other targets have included protein tyrosine phosphatases and matrix metalloproteinases.[41] It has also provided the means for the kinetically controlled capture of highly reactive minor abundance conformers of a fluctuating protein template.[42]

6.2.4 siRNA Approaches

Gene silencing is "switching off" a gene by a mechanism other than genetic modification and can occur naturally during the translational or post-translational stages. RNA interference (RNAi) uses small RNAs to suppress genes with complementary sequences thus preventing translation to deliver the gene product, usually a protein. A small interfering RNA (siRNA) (or short interfering RNA or silencing RNA), is a double-stranded RNA molecule about 20–25 nucleotides long which *inter alia* is involved in RNA interference to prevent expression of a specific gene.[45,46] The mechanism of SiRNA silencing is shown in Figure 6.5.

In principle, any gene of known sequence can be targeted by identifying an appropriate siRNA by screening siRNA libraries and this method has been used in gene function and drug target validation studies in the post-genomic era. Attempts are also being made to pursue the therapeutic potential of specific

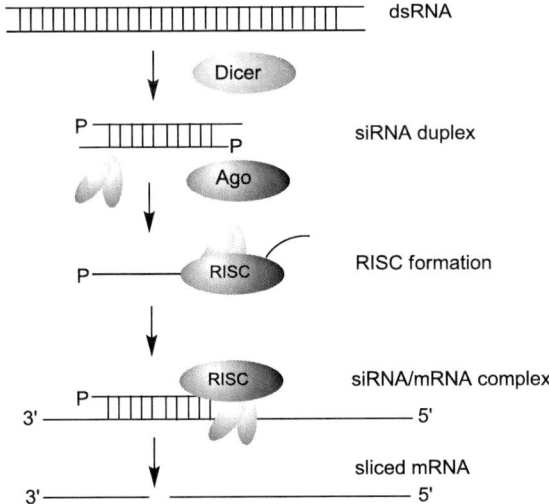

Figure 6.5 Mechanism of siRNA silencing: the RNAse III nuclease "Dicer" processes double-stranded RNA (dsRNA) and short hairpin RNA (shRNA) into short double-stranded RNA fragments (siRNA) of about 20–25 nucleotides long, usually with a two-base overhang on the 3′ ends. Dicer catalyses and initiates the formation of the RNA-induced silencing complex (RISC). During RISC assembly, siRNAs unwind and a single strand of RNA remains bound to the RISC. The RISC catalytic endonuclease component "Agonaute" cleaves the target messenger RNA (mRNA) in the middle of the region of homology to the siRNA thus preventing translation.

gene silencing. Two strategies for introducing small RNAs into the cell cyto-plasm that have been well explored are transfection, *i.e.* administering double-stranded RNAs in complexes designed to assist intracellular delivery, and gene therapy approaches to express precursor RNAs *in situ* from an appropriate vector, *e.g.* a plasmid, which is not a topic for this chapter.

Transfection methods vary. The most common is to administer siRNA as part of a complex with lipids and surfactants, optionally encapsulated within nanoparticles or liposomes, and there is a plethora of proprietary transfection products reflecting the fact that the delivery system must be tailored to the nature of the actual payload and the biological target. These methods have been successful in demonstrating the effects of siRNA in easily accessible tissue and local siRNA delivery has shown benefit in small animal models involving the lung, vagina, subcutaneous tissue, muscle, eye and central nervous system; several candidates have reached early phase clinical trials.[47,48] There is probably room for improvement in these delivery systems, perhaps through the study of precisely structured transfection agents such as Gemini surfactants.[49]

Delivery to less accessible sites probably requires systemic administration and thus provides a significant challenge in delivering a non-drug-like active moiety and avoiding deleterious "off-target" actions, such as dose-related liver toxicity and activation of the interferon defence mechanism.[50] A systemic drug based on siRNA would also need to achieve uptake into the right compartment in cells at all sites of action and be able to escape from endosomes and lyso-somes. Control of these properties probably lies beyond what can be achieved with adjuvenant transfection mixtures and much work has been targeted at well-tailored cationic complexes or nanoparticles.[48]

Chemical modifications of siRNA have shown that short-term activity, off-target actions and long-term activity can be modulated by structural change but the results are dependent on cell type and target. For the blood clotting initiator factor in a HaCaT cell line for either strand chemical modification at the 5'-end by 2'-*O*-methylation or 2'-*O*-allylation or G/C transversions were tolerated to a greater degree than modifications close to the 3'-terminal.[51] A similar result was found in studies of the silencing of endogenous lamin A/C in human HeLa and mouse SW3T3 cells, where it was found that modifications close to the 3'-end of the antisense strand were the most potent in abolishing silencing activity and effects were greater in HeLa than SW3T3 cells.[52] In both studies RNase-pro-tective phosphothiolate modifications led to extended silencing activity but when 50% of phosphates were so modified toxicity was observed. A 2'-fluoro-pyrimidine modification was also found to extend duration.[52]

A more detailed study of the effects of 2'-substitution provides evidence that the effect is independent of sequence.[53] In this study the key modification was a 29-*O*-methyl ribosyl substitution at position-2 in the guide strand, which reduced the silencing of most off-target transcripts with complementarity to the seed region of the siRNA guide strand. This sharp positional dependence contrasted with the broader position dependence of base-pair substitutions within the seed region, suggesting a role for position-2 of the guide strand

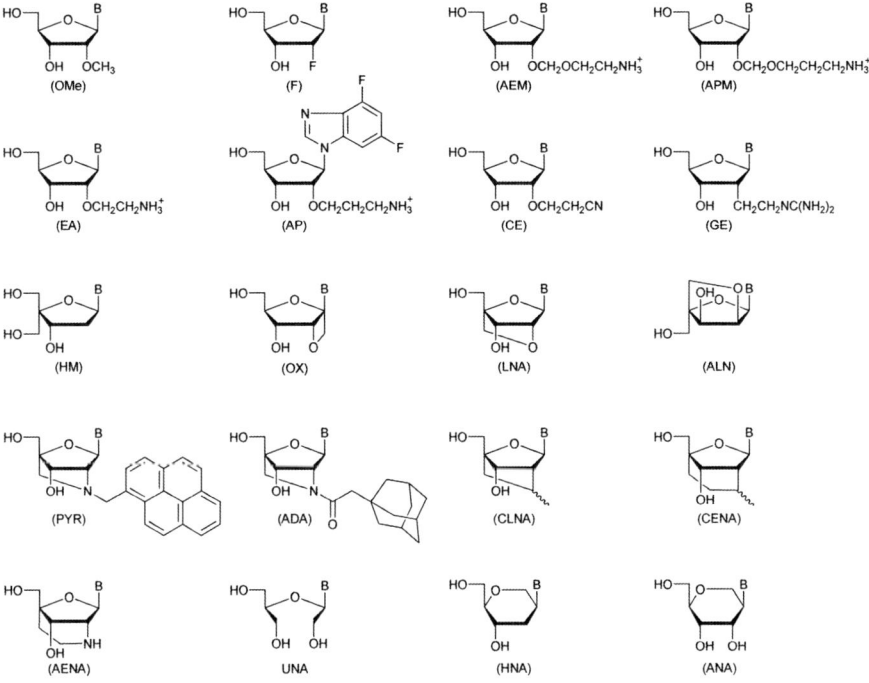

Figure 6.6 Reported ribose 2′-modifications present in siRNA duplexes.[54]

distinct from its effects on pairing to target transcripts.[53] A comparison of the effect of 21 types of chemical modifications (Figure 6.6) on siRNA activity and toxicity in a total of 2160 siRNA duplexes concludes that siRNA activity is primarily enhanced by favouring the incorporation of the intended antisense strand during RNA-induced silencing complex (RISC) loading by modulation of siRNA thermodynamic asymmetry and engineering of siRNA 3′-overhangs.[54]

Conjugation with lipophilic groups also enhanced cell uptake and improved pharmacokinetics and biodistribution in mice. An ApoB-specific siRNA with a chemically modified backbone conjugated to a 3′-lipophilic cholesterol moiety in the sense strand resulted in knockdown of ApoB mRNA by approximately 60% in the liver and approximately 75% in the jejunum, whilst no off-target effects or immune stimulation were noted despite the high 50 mg kg^{-1} dose needed for efficacy.[55]

Overall, the work in this area has been mainly limited to nucleotide mutations and simple derivatives of the ribose moiety,[54] and this has shown that SARs can be developed and that it may be possible to optimise selectivity and distribution. The way forward may be through more adventurous bioisosteric changes within the siRNA, perhaps to introduce moieties which have demonstrated mimicry of ribonucliotide function in *e.g.* ATP-dependent kinase inhibitors or cyclic nucleotide phosphodiesterases *etc.*

6.3   Knowledge-based Iterative Systems

6.3.1   Overview

Screening speculatively prepared or collected compounds for a chemical drug lead for a novel protein relies entirely on serendipity and can only be certain to find the universal "best" lead when the collection presents the entire drug-like universe. However, there is no way this level of representation can be achieved; current collections exemplify only an infinitesimal portion of the chemical universe. The introduction of the biological target into a dynamic combinatorial library, even with serial refinement of the reagents chosen, can only find the most appropriate lead within the scope of the library. In terms of chemical diversity this is probably even more limited than any sizable fixed collection. The siRNA approach seeks to change the game by targeting an achievable complete combinatorial library of components of a type known to be highly effective in binding to the final biological target. However, the type and number of targets is limited and there is a significant challenge for the medicinal chemist to effect sufficient, selective delivery of the warhead.

Knowledge-based lead discovery may be the only way of bypassing the restrictions in chemical diversity and target selection implicit in library-based approaches. This is almost the only method that involves a multi-step process of using information as it becomes available to update a notional pharmacophore (*i.e.* a summary of the position and nature of features perceivable by another molecule that either assist or hinder ligand binding at the protein target) thus describing the chemical properties that define a good lead and from which the structures of better leads might be predicted for synthesis. Of course, the process may not always be successful, but as it will always return instructive biological data there will always be a new understanding on which to base a new hypothesis. This cyclic process is iterated until a targeted biological profile is achieved. As it is new biological information relating to a specific target that is used to predict a new molecule, it is highly unlikely that the compound will be an undiscovered component of any pre-existing speculative library. In contrast to screening as a method of detecting leads, after several iterations of the knowledge based process it is highly unlikely that the final acceptable solution will have been one foreseen at the beginning of the search.

Outside the field of natural products, the pre-1990, high-value, novel blockbuster and breakthrough products were largely a product of iterative knowledge-based "rational design". The search for these molecules would be started from a seed structure or observation, or even *de novo* and the campaign would usually amount to the targeted synthesis of less than a thousand compounds for biological testing including the "non-rational" compounds generated to test hypotheses or limited serendipity excursions. However, it might have taken perhaps five or more years to produce what would now be seen as a relatively small number of compounds. This was because cyclical revision of synthetic targets usually yielded a predicted "improved" structure that lay in novel chemical space and required a lengthy exploration of novel synthetic routes. The slowness of pre-1990 iterative search, the weakness of the method as a way of addressing novel genomic targets where no lead will be known and the redirection of chemical resource to rapid expansion of screening collections all led to the abandonment of this method for divergent lead discovery. It is now mostly used in lead optimisation to converge on the development candidate.

The abandoning of iterative lead discovery through chemotype expansion was not a beneficial move with respect to lead quality. The unforeseen nature of its products inherently led to an exclusive chemical patent estate and the slow delivery of compounds provided enough time for extensive, multi-assay bio-logical profiling of every potential lead. Not only did this increase, relative to industrial methods, the breadth and quality of biological data to drive the design of new molecules over several relevant SARs, but occasionally this extended biology investigation would lead to the serendipitous discovery of an "off-target" novel therapeutic utility. The following sections aim to describe two of the current attempts to overcome the cost and slowness of the older process, whilst retaining the opportunities for unforeseen "black swan"[56] dis-coveries. The first describes the possible components of a strategically and economically acceptable version of the previously proven iterative lead dis-covery method. The second exploits more recent progress in the genomics area through a realisation system based on DNA-directed synthesis.

6.3.2 New Tools for Iterative Working

For the past two decades the development of discovery tools has focussed on the "zero cycle" screening paradigm. Current tools are mainly those that capitalise on serendipity and, based on the belief that a large enough compound collection will contain the best possible lead (or one sufficiently novel to justify market exclusivity), focus on the generation, management and screening of ever-increasing numbers of compounds. New synthesis hardware has targeted an industrialised process for producing and quality assuring compounds pre-pared *en masse* by a pre-ordained route. Accordingly cheminformatics software has been aimed mainly at providing the best spread of diversity over highest numbers of compounds derived using the smallest number of routes and reactions. It will usually also provide a management system for the large inventory of materials in process at any time. Downstream from the initial HTS, the hit-to-lead strategy is a serial triage of assays of reducing throughput and increasing stringency.

As almost the antithesis of the iterative process, little activity has been devoted to the provision of tools to enhance the iterative process and most of the improvements are simply generational upgrades of traditional manual equipment. The iterative process is unattractive to large product developers with existing HT-products and the development of tools for high throughput–high performance iterative discovery represents a significant challenge because chemical and biological targets are iteratively revised and a "run" cannot be pre-planned. Planning can extend no further than one of the many cycles necessary to realise the goal, and even within the cycle decisions may need to be made "on the fly" as circumstances arise. Therefore the synthesis apparatus must be adaptive and able to conduct route and reaction discovery. As far as possible, the assays are best conducted in parallel on a "per compound" basis with data being returned without delay to maximise the amount of multi-stream data available at the time of compound re-design. Ideally, any management software must support a multi-cycle real-time convergent search from any starting point without a synthetic preference or structural bias, which probably means it will need to supply "inspiration" at some point! However, the challenge has been taken up by a range of small and medium enterprise (SME) companies and academics and the first generation products are beginning to appear.

While it is an economic reality that the pharmaceutical industry cannot return to the slow manual pre-1990 iterative methodology for lead discovery, more recent history has shown that neither can it rely on serendipity alone for its good fortune. A way forward would seem to be lead discovery that takes the benefits of both processes. Thus, screening can supply a range of "hits" to act as seeds for an iterative lead discovery process, including hits for novel genomic proteins. The iterative step would drive lead expansion into novel areas using multi-stream assays as guides. This post-screening divergent path will increase novelty and thus patent class estate whilst reducing risk and increasing biological serendipity by identifying unforeseen orthogonal data-assured options. The challenge is to achieve this without extending cycle time or incurring new costs. Early work suggests that the solution may lie in downscaling the iterative process.

6.3.3 Fast and Efficient Realisation of Novel Leads: Rescaling the Process

The choke point in iterative lead expansion is the need to investigate novel routes to the novel compounds predicted to be better leads. This is particularly so when manual methods requiring tangible levels of compound (*i.e.* grams) are used. In turn this requires a replenishment relay of substantial amounts (up to hundreds of grams) of novel intermediates, often by an extended route. Emerging technologies in this area currently focus on the miniaturisation and integration of the chemical and biological processes with the aim of reducing reaction and work-up times and reducing material requirements. At their most efficient, the compound preparation is carried out at a scale that will produce only the "information" required rather than any storable amounts of material, where "information" could *e.g.* the identification of optimal conditions for a reaction, the best ordering of reactions to form a route, or the effect of a compound on a biological target.

Fluidic methods also ease the handling of small amounts of material and the management of an integrated process in which a compound generator module is linked directly to biological assay(s). Thus not only are the losses of time and materials associated with manual or robotic batch handling, bottling and transfer to another location avoided, but also the system is self-contained and independent of the corporate processes for managing and transferring compounds made *via* the "normal" larger scale process. The miniaturised process pin-points a select group of potential leads for larger scale synthesis.

Much of the equipment required to work at reduced scale had been available for some time but ultra-accurate low-volume pumps and robust glass chips and fittings required specific development. Because of the need to contain the high demand for protein in HTS, assays had been minimised and the amount of compound required to obtain a half-maximal response in many primary assays was down to nanograms. Micro-LC-MS methods for separation and identification were also optimal in this range, but the micro scale and its associated high surface to bulk ratios (about 10^{12} times greater than a macro flask) continued to set a challenge for synthesis and assay systems in terms of the materials, construction, and pump pressure limits. However, starting from the lead provided by electroosmotic uTAS systems,[57] non-ionic materials can be pressure-pumped in bonded glass chips with cross sectional dimensions in the range of a few microns and a working quantity of <1 microgram were developed.[58-62] These provide chemical robustness and avoid the reversible absorption of organic solvents and small molecule solutes which causes cross contamination in more easily fabricated plastic chips. Suitable glass chips, tubes and connectors with low dead volumes and high chemical resistance have become commercial items.

At the low micro-scale some additional benefits of miniaturisation are observed. Their "one-at-a-time" operational speed is reasonably matched to their mixing and reaction times (a few seconds) and many times faster than those of macro batch or flow systems. In addition, because of the reproducible heterogeneity of these low Reynolds number systems, a significant level of real-time optimisation can be achieved by modification of flows, porting points, applied charges *etc*. Thus "wait time", a significant factor at macro scale, is largely eliminated from the process and with suitable sensors and under appropriate software guidance the real-time adaptability of the system can support an element of goal-seeking to find the best possible output. This is achieved with higher certainty using less time and material than that required for the total number of larger scale pre-fixed batch experiments that would be needed to explore all of the possible conditions. Overall, this sets reagent and intermediate resources to a milligram level thus relieving many supply chain issues.

Early prototype systems, such as those shown in Figures 6.7 and 6.8 have demonstrated that the cycle time for conducting a chemistry reaction on a microgram scale can be as little as 10 s and that within a few minutes a concatenated flow assay system can take a small fraction of the output along with about 1/1000th the amount of protein normally found in one well of an HTS plate to deliver an IC_{50} value. Obviously there are many other possible

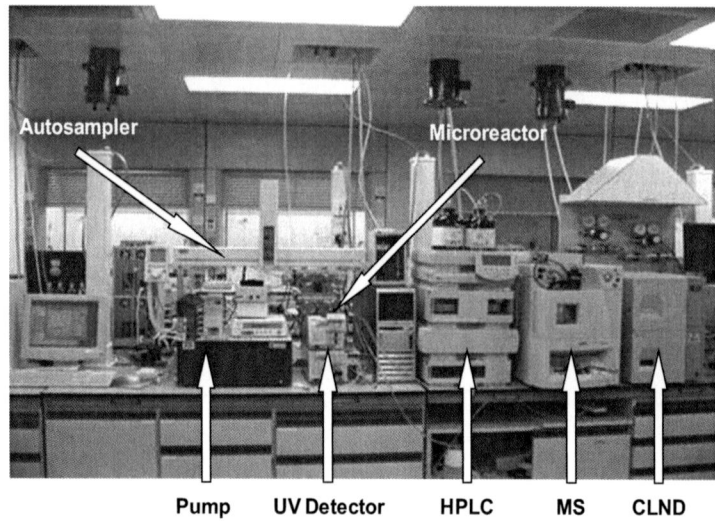

Figure 6.7   Chemistry generator unit closed loop prototype system (*ca.* 2004).

Figure 6.8   Example of a miniaturised flow assay device. The machine employs a 30 fL
sample volume and is sensitive down to around 400 molecules in the
confocal zone. A full IC_{50} curve requires 0.001% of the protein needed for
a single point in an HTS assay.

hardware arrangements, but what these systems demonstrate is that if only
information is required, then the time, manpower and intermediate resource
problems that plagued historic iterative chemistry can be avoided.

An early commercial exploitation of the very short reaction times found in micro-scaled chemistry is worth a specific mention as it has the potential to bring PET imaging technologies into the field of drug discovery where the ability to trace the distribution and fate of a drug substance is of higher value than imaging physiological changes. The fast synthesis times achievable in micro-scaled PET biomarker synthesis have already shifted the paradigm in the use of PET biomarkers in early R&D by enabling the ready synthesis of a variety of ^{18}F-labelled materials ($t_{1/2}$ = 120 min) at the point of use and reducing the need to identify special synthesis routes to insert a short-lived positron-emitting isotope in the final step.[63] With the additional promise of a bench-top micro-cyclotron by 2010,[64] the point-of-use production of experimental drugs containing short half-life isotopes such as ^{11}C ($t_{1/2}$ = 12 min) of virtually any micro-synthesisable molecule becomes a realistic goal[65] and from that a ready assessment of dynamic organ-time distribution and elimination.[66]

6.3.4 Prediction of Unforeseen Structures

Notionally, an iterative lead discovery system based on the hardware performance attributes described in Section 3.3 requires two types of decisions: synthetic chemistry decisions associated with finding a route to the predicted structure and medicinal chemistry decisions by which incoming and accumulated biological data are used to design a compound predicted to show an improved biological profile. If the speed and frugality of the hardware performance described in Section 3.3 is to be fully exploited to deliver an economically feasible lead diversification system, the time required for human data-driven decisions, particularly those requiring "inspiration", is a severe restriction. In addition, it is unlikely that most current computer-based modelling paradigms are appropriate for the task (see Figure 6.9).

Most medicinal chemistry modelling software requires a human to input a molecular structure. By this action the drawn structure is automatically something "foreseen" rather than novel and unexpected. These structure-based modelling systems have little or no ability in themselves to predict improved structures or even unforeseen bioisoteric fragment equivalents because nearly all evidence points to complete molecular models being holistic rather than the sum of fragments. In addition, as they employ "rules" based on former compounds they are essentially predictable. Historically, the iterative processes success lay on a human input, human idiosyncrasy, and human hunch and human inspiration, and the process was always greatly slowed by the capricious nature of these inputs. Software which seeks to emulate this behaviour is beginning to emerge and be taken up. One approach has been to automate the interpolation of the structural space between known active chemotypes.[67,68]

An alternative approach sets itself even further from any existing knowledge and tries to accelerate the way a human medicinal chemist "invents" new active structures, by using the power of the computer to make the process both faster and more reliable by considering a huge range of possible lead structures. These

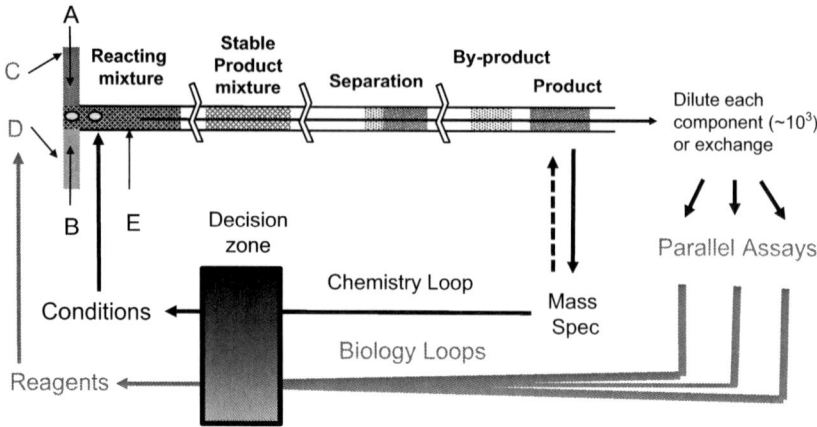

Figure 6.9   Schematic representation of a micro-scale (nanoflow) closed loop system. At very low scale and Reynolds number the nature, rate and physical arrangement of reagent inlet ports (A–E) can alter the nature of the output and using LC-monitoring the chemistry loop can exert a degree of real-time reaction optimisation. The decision zone indicates a point where human intervention (slow) or automated intervention (fast) must be applied.

methods call upon the notion of a lead pharmacophore rather than a lead "structure" as is usually presented as a cartoon depicting the connectivity of the underlying atoms. The working principles of an iteration are that the stored pharmacophore is updated by incoming biological data and this updated pharmacophore is used as the basis for predicting and ranking of a set of structures hypothesised to show improved biological performance. Thus in structural terms any compound, irrespective of chemotype, known or unknown, which has the potential to present or improve the running pharmacophore can be considered as a lead, thereby providing a system that is free from structural bias. Because of the minuscule amount of the drug-like universe currently known, the predicted set will almost certainly contain structures not foreseen before the run, because of the billions of *de novo* possibilities that can be examined in a short time. In addition, the running pharmacophore can manage the human working notions of "developability" by acting as a single container for a consensus of properties associated with activity at several biological targets (including those predicted by systems biology).

The most relevant type of pharmacophore is one that displays the properties present at the ligand surface as sensed by the protein, but this presents too severe a challenge to current computational speed for routine use. There are currently two approaches to more conveniently calculating and matching these molecular fields of a lead pharmacophore with the contents of huge databases of molecular structures and moieties *per se* or in trial assemblies. The most common solution is to make a basic shape/volume match and then use the small partial point charges placed at the centre of each atom in the basic molecular mechanics minimisation to generate an electrostatic overlay. A different approach uses an extended electron distribution (XED) model[69] which explicitly models the asymmetric distribution of electrons around atoms as its basic working element so that all important parameters can be matched simultaneously and uniquely, and the dynamic changes in field associated with conformational changes can be reflected. In a comparative study of force fields for a series of small molecules docked in pairs, XED force fields gave a better correlation ($r^2 = 0.96$) with experimental interaction energies than other well-known methods (MM2, 0.68; MM3, 0.68; OPLS, 0.42; AMBER, 0.17).[70] Field modelling is not limited to electrostatic descriptors and functions representing steric, hydrophobic and/or any other derivable properties which can also be used to generate fields (see Figure 6.10).

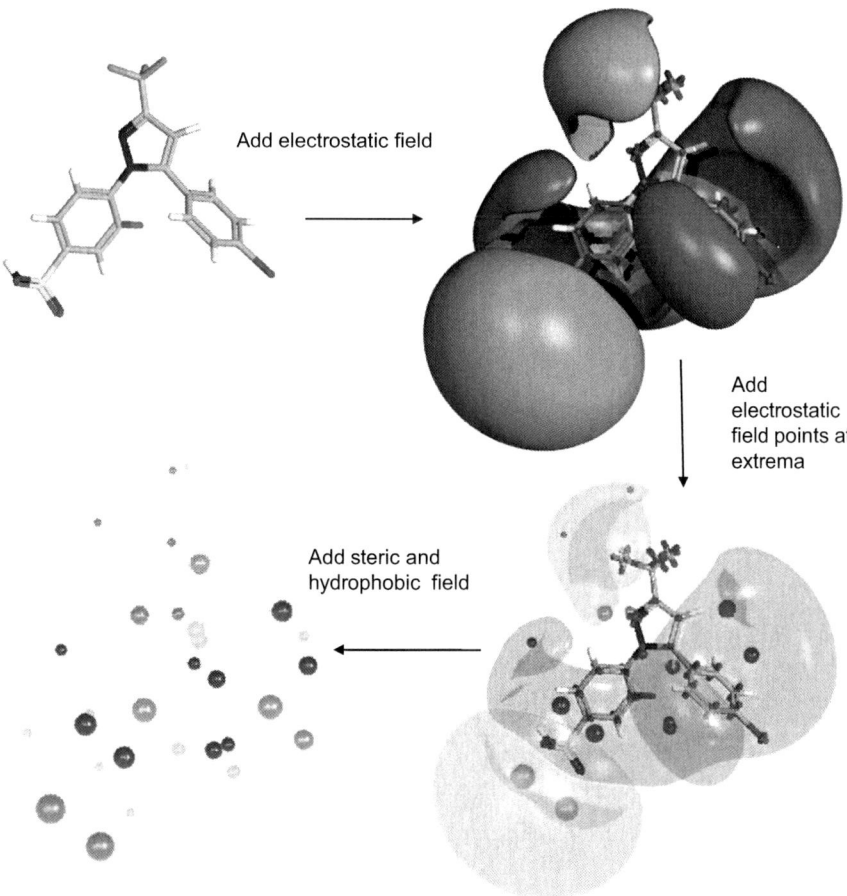

Add electrostatic field

Add electrostatic field points at extrema

Add steric and hydrophobic field

Figure 6.10 Translation of a standard structural cartoon to field points.

6.3.5   Challenges for High Speed Iterative Chemistry

Software systems to coordinate and manage automated chemistry and screening hardware are well developed. The next challenge is to understand how these systems can be developed to conduct a heuristic investigation of new routes to new compounds, as depicted by the "chemistry" loop of Figure 6.10. This is a significant challenge because it potentially encompasses the whole of synthetic organic chemistry. However, again the way forward may be to reflect how humans have accomplished this task through iterative methods. In this respect the use of the iterative method to focus and redefine the task may provide a way forward and the algorithms and philosophy described for prediction novel structures may have their counterparts in route screening. The "seeding" data can come from the historic data stored with the molecules and moieties of the design search database or from external retrosynthetic programs.

Miniaturised flow methods as described in Section 6.3.3 can provide the high speed and low costs needed to make a lead expansion phase economically sustainable within current pharmaceutical drug discovery strategies and a benefit of a sub-microgram product target is that for most synthons a 1 mg stock can be adequate as only the predicted compounds are synthesised, thus making local storage and fast retrieval of reagents feasible. Such a low requirement also greatly expands the range of compounds that can be made available as reagents. The bigger issue is what a reagent palette (or palettes) should contain. Similarly, the relationship between reaction repertoire and diversity range needs to be better understood. However, these problems are again no different from those faced by human operatives in the earlier era and the power of the iterative method is its ability to "learn" and constantly refocus its search and, with it, the "palette" of customised reagent and reactions needed. The initial impetus to a temporary lead could be provided by a systemised approach based on undemanding reactions and a limited reagent palette (*e.g.* click chemistry as represented in the extended scheme of Figure 6.4, or amide synthesis about a minimal scaffold.[71]

The management of an ongoing iterative route search is greatly assisted by the notion of synthetic "services", transformations that are essentially independent of the type of chemistry in play. Fortunately the range and facility of solid-supported reagents and catalysts as "services" in flow is rapidly expanding.[72] In addition, as the small scale of the experiment eliminates high energy hazards, small scale analogues of existing "services" for hydrogenation,[73] oxidation using either ozone or a new reagent $HOF \cdot MeCN$,[73,74] fluorination[75] and other reported flow transformations[76] should be possible.

The reduced scale and real-time optimisation capabilities of microscale synthesis systems operating at an "information only" level can also benefit from the utility of reactions normally discounted for large scale synthesis because of their exotic or toxic reagents, often used at stoichiometric levels. For example, C–H activation is a useful technique in the functionalisation of arenes and alkenes, but C–H activation in alkanes has been long known but under-exploited,[77,78] and may be more usefully explored by these small devices.

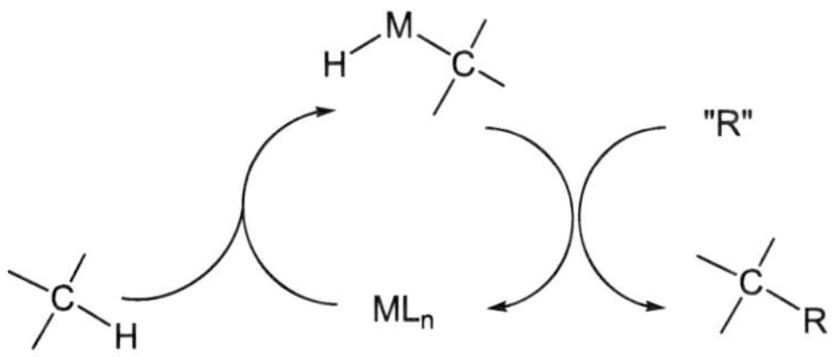

Figure 6.11 Schematic representation of alkane C–H activation using a metal catalyst.

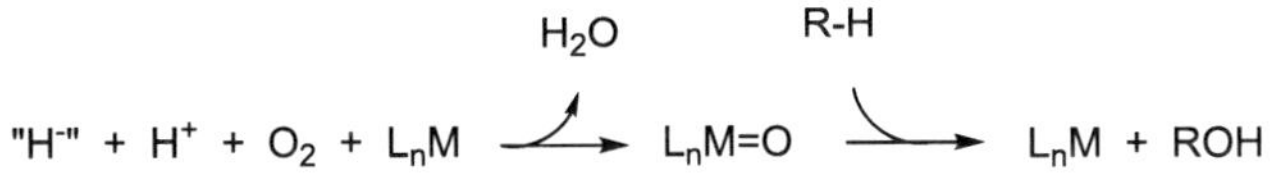

Figure 6.12 Monooxygenase process.

Formally, alkyl C–H activation requires insertion of a transition metal (usually Ru, Ir, Rh or Pd) across a strong C–H bond (90–105 kcal mol^{-1}) to form a new, weaker C–M bond (50–80 kcal mol^{-1}), followed by the generation of a new C–R bond (Figure 6.11). The onward reaction of the key intermediate with appropriate reagents "R" can lead *inter alia* to CO or isonitrile insertion, alkylation, borylation, 1,2-addition, sigma bond metathesis and (photo-catalysed) dehydrogenation. Extremely small systems can also employ Nature's C–H activation tools such as an isoform of cytochrome P-450[78] or methane monooxygenase[79] which perform a net insertion of an oxygen atom from dioxygen into a C–H bond. This is funded by NADH or another "H⁻"source to activate the O$_2$ molecule and the formation of water from one of the two O atoms, to provide the energy needed to break the O=O bond to give a high-energy high-oxidation state metal–oxo complex able to insert into the C–H bond (Figure 6.12).

6.3.6 Directed Assembly using DNA

DNA-Programmed Chemistry™ employs Watson–Crick base pairing to direct and enhance specific chemical reactions.[80,81] Dilute solutions of reactants (*ca.* nM concentrations) bearing suitable complementary DNA strands are annealed in very mild aqueous conditions thereby bringing the reactant moieties into close proximity and increasing their effective local concentration by a factor of $>10\,000$ (Figure 6.13). This has the effect of fixing their stoichiometry and greatly enhancing their rate of reaction. In some cases a modest level of stereo-selectivity has also been observed.[82]

This phenomenon facilitates a pre-programmed assembly of molecules using the basic principles illustrated in Figure 6.14 where the complementary

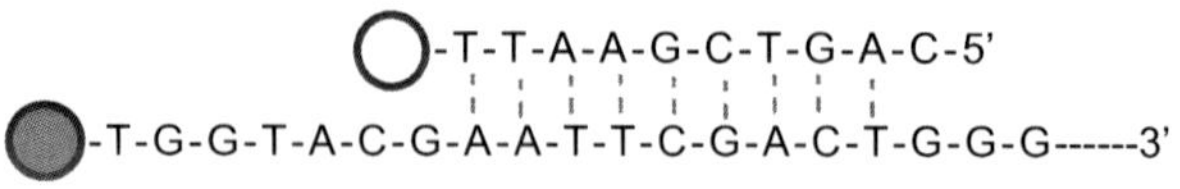

◯-T-T-A-A-G-C-T-G-A-C-5'
⬤-T-G-G-T-A-C-G-A-A-T-T-C-G-A-C-T-G-G-G------3'

Figure 6.13 Hybridisation of complementary DNA tags annealed to reactant moieties (represented by open and closed circles).

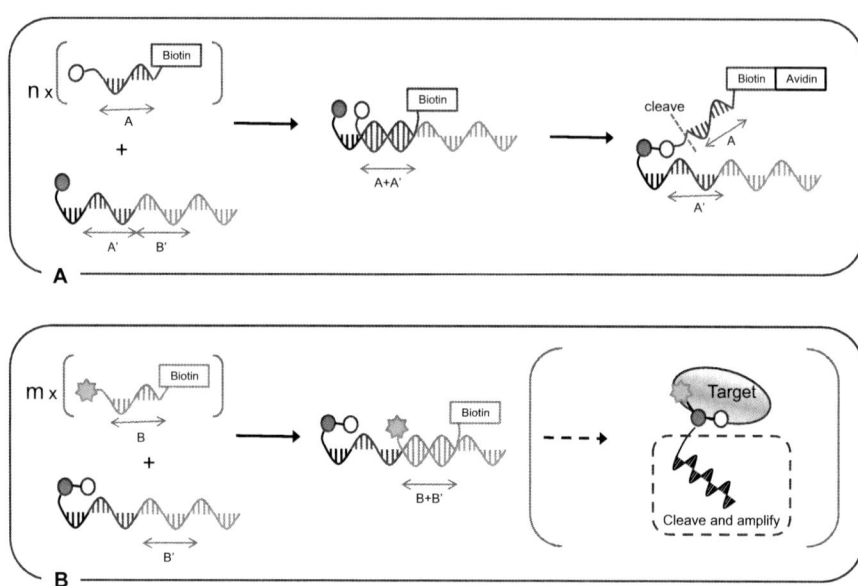

Figure 6.14 Cartoon representation of the key principles of DNA-Programmed Chemistry™.

components of reactants are colour-matched. For the case $n = 1$, Figure 6.14A shows a tagged reagent A (open circle) hybridising with the complementary shaded area A′ of the main tag attached to the blue reagent (closed circle). This then leads to the specific formation of the "red–blue" molecule A-A′. Several strategies have been used to assist product isolation; the use of biotin-based affinity capture and release is illustrated.[81] The commonly used cleavable linkers for tag and biotin attachment are based on sulfone, phosphane and diol chemistries. When $n = 1$, Figure 6.14B illustrates the principle of using a sequence of complementarities to drive the assembly of a specific molecule. Here reactant B (star) is introduced; and the code exists for a further "blue" reactant (not shown).

There are many ways that these basic working principles can be deployed and Li and Liu[81] summarise some elegant and inventive pathways. The relative position of the reacting moieties appears unimportant and it has been shown for some reactions that the apparent second-order rate constants did not change significantly when the distance between the hybridised reactive groups was varied over 30 base pairs.[83] A range of reactants and products unrelated to nucleic acids has been demonstrated including: S_N2 displacement,[83] conjugate addition,[83] reductive amination,[84] amine acylation,[84] oxazolidine formation,[85] nitro-aldol and nitro-Michael reactions,[84] Wittig olefination,[84] 1,3-nitrone cycloaddition,[84] Huisgen cycloaddition,[86] and Heck coupling reactions.[84]

The many-fold enhanced rate and degree of reaction the hybridised reactant pairs relative to non-hybridisable alternatives in the same reaction mixture allows the selection of one reaction in the presence of many possibilities. Thus, the effective molarity of hybridised pairs can be used to perform otherwise incompatible reactions simultaneously in the same solution. For example, it has been shown that three reactions of maleimides (amine addition, thiol addition, and nitro-Michael addition) are controlled to generate only the three sequence-programmed products out of the nine possible; two aldehyde couplings (reductive amination and Wittig olefination) can be carried out discretely in one solution, where three amine reactions (acylation, reductive amination, and maleimide addition) can be controlled to give only the DNA-templated products. In addition, all six reactions were performed simultaneously in the same solution to give only the 6 programmed products out of the 28 possibilities.[87]

Clearly from Figure 6.14 when n and/or $m > 1$, that is to say several reactants are coded with the same codon "colour", and there is no product isolation stage then combinatorial libraries can be formed. The process can be used to provide a solution mixture library of high component count, each component being present at very low concentration. This template-linked library can then be subjected to *in vitro* selection to identify members with the desired ligand properties.[88] The coding templates of library members selected can then be cleaved and amplified by PCR and either sequenced to identify desired compounds, or diversified and subjected to additional iterative cycles of DNA-targeted synthesis, selection, and amplification.[83]

This process could also be run to a different outcome: reaction discovery. DNA-templated organic synthesis and *in vitro* selection are used to simultaneously evaluate many combinations of different substrates for bond-forming reactions in a single solution. Bond-forming substrate combinations are then revealed using *in vitro* selection for bond formation, PCR amplification and DNA microarray analysis.[89]

Overall, DNA-programmed chemistry fulfils many of the requirements of a discovery system. It can produce discovery libraries in a combinatorial fashion to yield original hits and it can perform iterative loops based on recovery and amplification of the codes associated with successful ligands in *in vitro* selection filters. Following some sort of SAR analysis, the synthesis of predicted improved leads can be coded. The main limitation is that the chemistry must not affect the integrity of the DNA template. However, the power to discover reactions may offset this restriction.

6.4   Conclusion

With the current industrial screening-based lead-discovery paradigm under question, this has been an appropriate time to consider the alternatives. In the industrial model in the absence of any timely feedback to drive was towards high numbers of compounds to increase serendipity. However, without feedback it is also difficult to find an exceptional novel solution. The long term quality of compounds prepared by the industrial method during the 1990's can now be assessed by considering their attrition rate and the cause of attrition. As many chemists suspected, lead "quality", a term that may approximate to "novelty", was impaired relative to the preceding iterative era and this has had a significant negative impact on approval rates and exclusivity.

This chapter has attempted to critically assess some of the emerging strategies and technologies for iterative methods which, with full development, could be the methods of the future. No method has yet emerged which fully equates to the ideal of unrestricted guided lead discovery, the key prerequisite of finding a novel lead that exceeds historic knowledge. The chemical boundaries of some target-guided methods are inherent, either because of the restricted range of chemistry that can be employed (*e.g.* dynamic libraries) or because the protein target or the biological target must survive contact with the reaction mixture (or both). Clearly, isolation and re-dissolution of the test compound avoids this problem but on a manual scale this has proved to be a too lengthy and resource-intensive process to be viable, even starting from screened leads to provide a starting point for lead-less targets. Therefore the miniaturised integrated chemistry–biology hardware currently being explored as a way of delivering very short iterative cycle-times and very low resource demands have been reviewed. In addition, the question of how these ultra-fast systems can be controlled and guided with the human scientist being the supervisor but never the limiter of speed has been considered.

Whilst the basic hardware and software of a practical micro-scale system has been described, much work has yet to be done to understand the chemistry strategies and reagent palettes that are needed to direct and supply these fast integrated chemistry and biology systems. Two different approaches have been examined. DNA-directed chemistry incorporates a container, a DNA strand to which synthesis commands can be written for execution and from which reports can be received, thus providing the practical underpinnings for knowledge based iteration, but only at the expense of chemical scope. The other approach assumes that all exploratory chemistry will be carried out in a miniaturised environment, and the aim is to provide an appropriate guidance system for these ultra-fast reaction and transfer regimes. It is likely that the philosophy behind each approach will permeate the other to provide a new way to go. The good news is that a fully implemented fast iterative engine may not actually need to travel very far. A fact easily overlooked in the metrics-obsessed world of modern medicinal chemistry is that most *iterative* campaigns achieved lasting market success based on less than 1000 tested molecules. For example, Sir James Black's team synthesised only 700 compounds to discover the original

Histamine H2-receptor agonist burimamide and the total was still less than 1000 to achieve the final marketable product cimetidine, which enjoyed world-beating profitability during its 5 years of market exclusivity.[90,91]

References

1. US Food and Drug Administration, *Challenge and opportunity on the critical path to new medical products*, 2007, available from http://www.fda.gov/oc/initiatives/criticalpath/whitepaper/html.
2. J. Owens, *Nat. Rev. Drug Discovery*, 2007, **6**, 99.
3. A. Biancardi and S. Green, *Scrip*, 2008, **100**, S33.
4. D. Jackson, *The Pharmaceutical Market Outlook to 2010*, Business Insights, London, 2003.
5. A. Witty, CEO GSK, Reuters 19 July, 2009: "We've really thrown into reverse much of the trend of research organisation that had developed over the last 15 years. **Over that time, the drugs industry was a big commercial success but it took a "wrong turn" by deciding that drug discovery was an industrial process based on large-scale application of technologies like genomics, proteomics and combinatorial chemistry.** These were all supposed to transform productivity yet none of them did." Available from http://www.reuters.com/article/rbssPharmaceuticals%20-%20Generic%20&%20Specialty/idUSLJ9191220090619.
6. R. S. Bohacek, C. McMartin and W. C. Guida, *Med. Res. Rev.*, 1996, **16**, 3.
7. D. Brown, *Mol. Diversity*, 1996, **2**, 217.
8. A. Furka, F. Sebestyen, M. Asgedom and G. Dibo, *Int. J. Protein Res.*, 1991, **37**, 487.
9. F. Sebestyen, G. Dibo, A. Kovacs and A. Furka, *Bioorg. Med. Chem. Lett.*, 1993, **3**, 413.
10. K. S. Lam, S. E. Salmon, E. M. Hersh, V. J. Hruby, W. M. Kamiersky and R. J. Knapp, *Nature*, 1991, **354**, 82.
11. R. A. Houghten, C. Pinilla, S. E. Blondelle, J. R. Appel, C. T. Dooley and J. H. Cuervo, *Nature*, 1991, **354**, 84.
12. J.-M. Lehn, *Dynamic Combinatorial Chemistry and Virtual Combinatorial Libraries*, in *Essays in Contemporary Chemistry: from Molecular Structure towards Biology*, ed. G. Quinkert and M. V. Kisakürek, Verlag Helvetica Chemica Acta, Zürich, 2001, p. 307.
13. J.-M. Lehn, *Supramolecular Chemistry. Concepts and Perspectives*, VCH, Weinheim, 1995.
14. S. Otto, R. L. E. Furlán and J. K. M. Sanders, *Drug Discovery Today*, 2002, **7**, 117–125.
15. O. Ramström and J.-M. Lehn, *ChemBioChem*, 2000, **1**, 41.
16. I. Huc and J.-M. Lehn, *Proc. Natl. Acad. Sci. U. S. A.*, 1997, **94**, 2106.
17. G. R. L. Cousins, S.-A. Poulsen and J. K. M. Sanders, *Chem. Commun.*, 1999, 1575.

18. V. A. Polyakov, M. I. Nelen, N. Nazarpack-Kandlousy, A. D. Ryabov and A. V. Eliseev, *J. Phys. Org. Chem.*, 1999, **12**, 357.
19. S. J. Rowan and J. K. M. Sanders, *J. Org. Chem.*, 1998, **63**, 1536.
20. C. Amatore, A. Jutand, G. Meyer and L. Mottier, *Chem.–Eur. J.*, 1999, **5**, 466.
21. P. G. Swann, R. A Casanova, A. Desai, M. M. Frauenhoff, M. Urbancic, U. Slomczynska, A. J. Hopfinger, G. C. Lebreton and D. Venton, *Biopolymers*, 1996, **40**, 617.
22. D. M. Lynn, S. Kanaoka and R. H. Grubbs, *J. Am. Chem. Soc.*, 1996, **118**, 784.
23. R. H. Grubbs, S. J. Miller and G. C. Fu, *Acc. Chem. Res.*, 1995, **28**, 446.
24. C. Brändli and T. R. Ward, *Helv. Chim. Acta*, 1998, **81**, 1616.
25. V. Goral, M. I. Nelen, A. V. Eliseev and J.-M. Lehn, *Proc. Natl. Acad. Sci. U. S. A.*, 2001, **98**, 1347.
26. J. Leclaire, L. Vial, S. Otto and J. K. M. Sanders, *Chem. Commun.*, 2005, 1959.
27. Z. Rodriguez-Docampo and S. Otto, *Chem. Commun.*, 2008, 5301–5303.
28. K. C. Nicolaou, R. Hughes, S. Y. Cho, N. Winssinger, C. Smethurst, H. Labischinski and R. Endermann, *Angew. Chem., Int. Ed.*, 2000, **39**, 3823.
29. D. J. Maly, I. C. Choong and J. A. Ellman, *Proc. Natl. Acad. Sci. U. S. A.*, 2000, **97**, 2419.
30. H. C. Kolb, M. G. Finn and K. B. Sharpless, *Angew. Chem., Int. Ed.*, 2001, **40**, 2004.
31. R. A. Evans, *Aust. J. Chem.*, 2007, **60**, 384.
32. R. Huisgen, *Angew. Chem., Int. Ed. Engl.*, 1963, **2**, 633.
33. C. W. Tornoe, C. Christensen and M. Meldal, *J. Org. Chem.*, 2002, **67**, 3057.
34. L. Zhang, X. Chem, P. Xue, H. H. Sun, I. D. Williams, K. B. Sharpless, V. V. Fokin and G. Jia, *J. Am. Chem. Soc.*, 2005, **127**, 15998.
35. Y. Wang, W.-Y. Lin, K. Liu, R. J. Lin, M. Selke, H. C. Kolb, N. Zhang, X.-Z. Zhao, M. E. Phelps, C. K. F. Shen, K. F. Faull and H.-R. Tseng, *Lab Chip*, 2009, **9**, 2281.
36. J. Waser, B. Gaspar, H. Nambu and E. M. Carreira, *J. Am. Chem. Soc.*, 2006, **128**, 11693.
37. O. Ramström, S. Lohmann, T. Bunyapaiboonsri and J.-M. Lehn, *Chem.–Eur. J.*, 2004, **10**, 1711.
38. J. W. Kehoe, D. J. Maly, D. E. Verdugo, J. I. Armstrong, B. N. Cook, Y.-B. Ouyang, K. L. Moore, J. A. Ellman and C. R. Bertozzi, *Bioorg. Med. Chem. Lett.*, 2002, **12**, 329.
39. W. G. Lewis, L. G. Green, F. Grynszpan, Z. Radiæ, P. R. Carlier, P. Taylor, M. G. Finn and K. B. Sharpless, *Angew. Chem., Int. Ed.*, 2002, **41**, 1053.
40. R. Srinivasan, J. Li, S. L. Ng, K. A. Kalesh and S. Q. Yao, *Nat. Protoc.*, 2007, **2**, 2655.
41. Y. Bourne, H. C. Kolb, Z. Radić, K. B. Sharpless, P. Taylor and P. Marchot, *Proc. Natl. Acad, Sci. U. S. A.*, 2004, **101**, 1449.

42. R. Manetsch, A. Krasinski, Z. Radić, J. Raushel, P. Taylor, K. B. Sharpless and H. C. Kolb, *J. Am. Chem. Soc.*, 2004, **126**, 12809.
43. A. Krasiñski, R. Radić, J. Manetsch, J. Raushel, P. Taylor, K. B. Sharpless and H. C. Kolb, *J. Am. Chem. Soc.*, 2005, **127**, 6686.
44. V. P. Mocharla, B. Colasson, L. V. Lee, S. Röper, K. B. Sharpless, C.-H. Wong and H. C. Kolb, *Angew. Chem., Int. Ed.*, 2005, **44**, 116.
45. S. Elbashir, J. Harborth, W. Lendeckel, A. Yalcin, K. Weber and T. Tuschl, *Nature*, 2001, **411**, 494.
46. A. Hamilton and D. Baulcombe, *Science*, 1999, **286**, 950.
47. D. M. Dykxhoorn, D. Palliser and J. Lieberman, *Gene Ther.*, 2006, **13**, 541.
48. S. Akhtar and I. F. Benter, *J. Clin. Invest.*, 2007, **117**, 3623.
49. A. J. Kirby, P. Camilleri, J. B. F. N. Engberts, M. C. Feiters, R. J. M. Nolte, O. Soderman, M. Bergsma, P. C. Bell, M. L. Fielden, C. L. Garcia-Rodriguez, P. Guedat, A. Kremer, C. McGregor, C. Perrin, G. Ronsin and M. C. P. van Eijk, *Angew. Chem., Int. Ed.*, 2003, **42**, 1448.
50. S. Frantz, *Nat. Rev. Drug Discovery*, 2006, **5**, 528.
51. M. Amarzguioui, T. Holen, E. Babaie and H. Prydz, *Nucleic Acids Res.*, 2003, **31**, 589.
52. J. Harborth, S. M. Elbashir, K. Vandenburgh, H. Manninga, S. A. Scaringe, K. Weber and T. Tuschl, *Antisense Nucleic Acid Drug Dev.*, 2003, **13**, 83.
53. A. L. Jackson, J. Burchard, D. Leake, A. Reynolds, J. Schelter, J. Guo, J. M. Johnson, L. Lim, J. Karpilow, K. Nichols, W. Marshall, A. Khvorova and P. S. Linsley, *RNA*, 2006, **12**, 1197.
54. J. B. Bramsen, M. B. Laursen, A. F. Nielsen, T. B. Hansen, C. Bus, N. Langkjær, B. R. Babu, T. Højland, M. Abramov, A. Van Aerschot, D. Odadzic, R. Smicius, J. Haas, C. Andree, J. Barman, M. Wenska, P. Srivastava, C. Zhou, D. Honcharenko, S. Hess, E. Muller, G. V. Bobkov, S. N. Mikhailov, E. Fava, T. F. Meyer, J. Chattopadhyaya, M. Zerial, J. W. Engels, P. Herdewijn, J. Wengel and J. Kjems, *Nucleic Acids Res.*, 2009, **37**, 2867.
55. J. Soutschek, A. Akinc, B. Bramlage, K. Charisse, R. Constien, M. Donoghue, S. Elbashir, A. Geick, P. Hadwiger, J. Harborth, M. John, V. Kesavan, G. Lavine, R. K. Pandey, T. Racie, K. G. Rajeev, I. Röhl, I. Toudjarska, G. Wang, S. Wuschko, D. Bumcrot, V. Koteliansky, S. Limmer, M. Manoharan and H.-P. Vornlocher, *Nature*, 2004, **432**, 173.
56. N. N. Taleb, *The Black Swan: The Impact of the Highly Improbable*, Random House Inc, New York, 2007.
57. P. D. I. Fletcher, S. J. Haswell and V. N. Paunov, *Analyst*, 1999, **124**, 1273.
58. V. Skelton, G. M. Greenway, S. J. Haswell, P. Styring, D. O. Morgan, B. H. Warrington and S. Y. F. Wong, *Analyst*, 2001, **126**, 7.
59. V. Skelton, G. M. Greenway, S. J. Haswell, P. Styring, D. O. Morgan, B. H. Warrington and S. Y. F. Wong, *Analyst*, 2001, **126**, 11.
60. P. D. I. Fletcher, S. J. Haswell, E. Pombo-Villar, B. H. Warrington, P. Watts, S. Y. F. Wong and X. L. Zhang, *Tetrahedron*, 2002, **58**, 4735.
61. R. Mukhopadhyay, *Anal. Chem.*, 2007, **79**, 3248.

62. S. Y. F. Wong-Hawkes, J. C. Matteo, B. H. Warrington and J. D. White, in *Microreactors as New Tools for Drug Discovery and Development in New Avenues to Efficient Chemical Synthesis Emerging Technologies*, ed. P. H. Seeberger, T. Blume, Springer-Verlag, Heidelberg, 2007, p. 39.
63. R. Nutt, L. J. Vento and M. H. T. Ridinger, *Clin. Pharmacol. Ther.*, 2007, **81**, 792.
64. See http://www.advancedbiomarker.com/
65. A. Smith, D. Patton, A. Giamis and J. C. Matteo, *J. Nucl. Med.*, 2008, **49**(Suppl. 1), 46P.
66. L. A. Green, S. S. Gambhir, A. Srinivasan, P. K. Banerjee, C. K. Hoh, S. R. Cherry, S. Sharfstein, J. R. Barrio, H. R. Herschman and M. E. Phelps, *J. Nucl. Med.*, 1998, **39**, 729.
67. S. Wetzel, K. Klein, S. Renner, D. Rauh, T. I. Oprea, P. Mutzel and H. Waldmann, *Nat. Chem. Biol.*, 2009, **5**, 581.
68. S. Renner, W. A. L. van Otterlo, M. D. Seoane, S. Möcklinghoff, B. Hofmann, S. Wetzel, A. Schuffenhauer, P. Ertl, T. I. Oprea, D. Steinhilber, L. Brunsveld, D. Rauh and H. Waldmann, *Nat. Chem. Biol.*, 2009, **5**, 585.
69. J. G. Vinter, *J. Comput.-Aided Mol. Des.*, 1994, **8**, 653.
70. J. G. Chessari, C. A. Hunter, C. M. R. Low, M. J. Packer, J. G. Vinter and C. Zonta, *Chem.–Eur. J.*, 2002, **8**, 2860.
71. R. D. Cramer, D. E. Patterson, R. D. Clark, F. Soltanshahi and M. S. Lawless, *J. Chem. Inf. Comput. Sci.*, 1998, **38**, 1010.
72. For example see http://leygroup.ch.cam.ac.uk/publications.html.
73. I. Kovacs, R. Jones, Z. Otvos, L. Urge, G. Dorman and F. Darvas, in *Heterogeneous Catalysis Research Progress*, ed. M. B. Gunther, Nova Science Publisher, Inc., New York, 2008, p. 395.
74. C. B. McPake, C. B. Murray and G. Sandford, *Tetrahedron Lett.*, 2009, **50**, 1674.
75. R. D. Chambers, D. Holling, G. Sandford, A. S. Batsanov and J. A. K. Howard, *J. Fluorine Chem.*, 2004, **125**, 661.
76. W. Ehrfeld, V. Hessel and H. Lowe, *Microreactors. New Technology for Modern Chemistry*, Wiley-VCH, New York, 2000.
77. A. S. Goldman and K. I. Goldberg, Organometallic C–H Bond Activation: An Introduction, in *Activation and Functionalization of C–H Bonds*, ACS Symposium Series, *2004*, **885**, 1.
78. B. Meunier, *Biomimetic Oxidations Catalyzed by Transition Metal Complexes*, Imperial College Press, River Edge, NJ, 2000.
79. M.-H. Baik, M. Newcomb, R. A. Friesner and S. J. Lippard, *Chem. Rev.*, 2003, **103**, 2385.
80. Z. J. Gartner, M. W. Kanan and D. R. Liu, *J. Am. Chem. Soc.*, 2002, **124**, 10304.
81. X. Li and D. R. Liu, *Angew. Chem., Int. Ed.*, 2004, **43**, 4848.
82. X. Li and D. R. Liu, *J. Am. Chem. Soc.*, 2003, **125**, 10188.
83. Z. J. Gartner and D. R. Liu, *J. Am. Chem. Soc.*, 2001, **23**, 6961.
84. Z. J. Gartner, M. W. Kanan and D. R. Liu, *Angew. Chem., Int. Ed.*, 2002, **41**, 1796.

85. X. Li, Z. J. Gartner, B. N. Tse and D. R. Liu, *J. Am. Chem. Soc.*, 2004, **126**, 5090.
86. Z. J. Gartner, R. Grubina, C. T. Calderone and D. R. Liu, *Angew. Chem., Int. Ed.*, 2003, **42**, 1370.
87. C. T. Calderone, J. W. Puckett, Z. J. Gartner and D. R. Liu, *Angew. Chem., Int. Ed.*, 2002, **41**, 4104.
88. Z. J. Gartner, B. N. Tse, R. Grubina, J. B. Doyon, T. M. Snyder and D. R. Liu, *Science.*, 2004, **305**, 1601.
89. M. W. Kanan, M. M. Rozenman, K. Sakurai, T. M. Snyder and D. R. Liu, *Nature*, 2004, **431**, 545.
90. J. W. Black, W. A. M. Duncan, G. J. Durant, C. R. Ganellin and M. E. Parsons, *Nature*, 1972, **236**, 385.
91. M. E. Parsons and C. R. Ganellin, *Brit. J. Pharmacol.*, 2009, **147**, S127.

Subject Index

References to figures, schemes or tables are given in *italic* type